ANSI/ISA-S5.1-1984
Approved November 5, 1986

American National Standard

Instrumentation Symbols and Identification

Instrument Society of America

ISBN 0-87664-844-8

ISA-S5.1 Instrumentation Symbols and Identification

INSTRUMENT SOCIETY OF AMERICA
67 Alexander Drive
P.O. Box 12277
Research Triangle Park, North Carolina 27709

PREFACE

This preface is included for information and is not a part of ISA-S5.1.

This standard has been prepared as part of the service of the Instrument Society of America (ISA) toward a goal of uniformity in the field of instrumentation. To be of real value, this document should not be static, but should be subject to periodic review. Toward this end, the Society welcomes all comments and criticisms, and asks that they be addressed to the Secretary, Standards and Practices Board, Instrument Society of America, 67 Alexander Drive, P.O. Box 12277, Research Triangle Park, NC 27709, Telephone (919) 549-8411.

The ISA Standards and Practices Department is aware of the growing need for attention to the metric system of units in general, and the International System of Units (SI) in particular, in the preparation of instrumentation standards. The Department is further aware of the benefits to U.S.A. users of ISA standards of incorporating suitable references to the SI (and the metric system) in their business and professional dealings with other countries. Toward this end, this Department will endeavor to introduce SI-acceptable metric units in all new and revised standards to the greatest extent possible. The Metric Practice Guide, which has been published by the Institute of Electrical and Electronics Engineers as ANSI/IEEE Std. 268-1982, and future revisions will be the reference guide for definitions, symbols, abbreviations, and conversion factors.

It is the policy of the Instrument Society of America to encourage and welcome the participation of all concerned individuals and interests in the development of ISA standards. Participation in the ISA standards-making process by an individual in no way constitutes endorsement by the employer of that individual, of the Instrument Society of America, or of any of the standards that ISA develops.

The information contained in the preface, footnotes, and appendices is included for information only and is not a part of the standard.

The instrumentation symbolism and identification techniques described in the standard accommodate the advances in technology and reflect the collective industrial experience gained since the publication of Recommended Practice RP5.1 in 1949.

This revision attempts to strengthen the standard in its role as a tool of communication in the process industries. Communication presupposes a common language; or, at the very least, it is facilitated by one. The standard offers the foundation for that common language.

When integrated into a system, the symbols and designations presented here form a concise, dedicated language which communicates concepts, facts, intent, instructions, and knowledge about measurement and control systems in the process industries.

This document is a consensus standard rather than a mandatory one. As such, it has many of the strengths and the weaknesses of consensus standards. Its primary strength is that it can be used in widespread, interdisciplinary ways. Its weakness is generally that of not being specific enough to satisfy the special requirements of particular interest groups.

The symbols and identification contained in ISA-S5.1 have evolved by the consensus method and are intended for wide application throughout the process industries. The symbols and designations are used as conceptualizing aids, as design tools, as teaching devices, and as a concise and specific means of communication on all types and kinds of technical, engineering, procurement, construction, and maintenance documents.

In the past, the standard has been flexible enough to serve all of the uses just described. In the future, it must continue to do so. To this end, this revision offers symbols, identification, and definitions for concepts that were not previously described; for example, shared display/control, distributed control, and programmable control. Definitions were broadened to accommodate the fact that, although similar functions are being performed by the new control systems, these functions are frequently not related to a uniquely identifiable instrument; yet they still must be conceptualized and identified. The excellent SAMA (Scientific Apparatus Makers Association) method of functional diagramming was used to describe function blocks and function designators. To help the batch processing industries, where binary (on-off) symbolism is extremely useful, new binary line symbols were introduced and first-letter *Y* was selected to represent an initiating variable which could be categorized as an event, presence, or state. In general, breadth of application as opposed to narrowness has been emphasized.

The ISA Standards Committee on Instrumentation Symbols and Identification operates within the ISA Standards and Practices Department, with William Calder III as vice president. The persons listed below served as members of or advisors to the SP5.1 committee. The SP5.1 committee is deeply appreciative of the work of previous SP5.1 committees and has tried to treat their work with the respect it deserves. In addition, this committee would like to acknowledge the work of the SP5.3 committee in developing ISA-S5.3, ''Graphic Symbols for Distributed Control/ Shared Display Instrumentation, Logic and Computer Systems.'' The key elements of ISA-S5.3 have been incorporated into ISA-S5.1, and it is the Society's intent to withdraw ISA-S5.3 after publication of this revision of ISA-S5.1.

The following people served as members of ISA Committee SP5.1, which prepared this standard:

NAME	COMPANY
R. Mulley, Chairman	Fluor Engineers, Inc.
E. J. Blahut	Blahut Engineering, Inc.
P. R. Boubel	TXE Inc.
J. P. Carew	Stone and Webster Engineering Corporation
N. Dogra	ANK Engineers
J. E. Doyle	Tweedcrest Limited
C. R. Gross	EXXON Company U.S.A.
T. E. Hamler	Owens Corning Fiberglas Corporation
F. Horn	Allied Chemical Company
A. A. Iverson	ARCO Chemical Company
A. Langelier	Polaroid Coroporation
W. E. Mapes	Eastman Kodak Company
T. C. McAvinew	Vertech Treatment Systems
W. L. Mostia	AMOCO Chemicals
G. K. Pace	Phelps Dodge Corporation
*G. Platt, Past Chairman	Bechtel Power Corporation
A. W. Reeve	AWR Controls (Canada) Ltd.
S. Sankaran	McDermott Engineering
R. M. Shah	Olin Chemicals Corporation
D. G. Turnbull	Sandwell and Company, Limited
R. von Brecht	The M. W. Kellogg Company
G. Wilbanks	The Rust Engineering Company

*Member Emeritus

The following people served as members of ISA Committee SP5:

NAME	COMPANY
D. E. Rapley, Chairman	Stearns Catalytic Corporation
R. C. Greer	Bailey Controls Company
D. G. Kempfer	Standard Oil Company of Ohio
R. H. Kind	El Paso Natural Gas Company
R. Mulley	Fluor Engineers, Inc.
T. J. Myron	The Foxboro Company

This standard was approved for publication by the ISA Standards and Practices Board in September 1984.

NAME	COMPANY
W. Calder III, Chairman	The Foxboro Company
P. V. Bhat	Monsanto Company
N. L. Conger	Conoco
B. Feikle	Bailey Controls Company
H. S. Hopkins	Westinghouse Electric Company
J. L. Howard	Boeing Aerospace Company
R. T. Jones	Philadelphia Electric Company
R. Keller	The Boeing Company
O. P. Lovett, Jr.	ISIS Corporation
E. C. Magison	Honeywell, Inc.
A. P. McCauley	Chagrin Valley Controls, Inc.
J. W. Mock	Bechtel Corporation
E. M. Nesvig	ERDCO Engineering Corporation
R. Prescott	Moore Products Company
D. E. Rapley	Stearns Catalytic Corporation
W. C. Weidman	Gilbert Commonwealth, Inc.
K. A. Whitman	Consultant
*P. Bliss	Consultant
*B. A. Christensen	Contintental Oil Company
*L. N. Combs	Retired
*R. L. Galley	Consultant
*T. J. Harrison	IBM Corporation
*R. G. Marvin	Roy G. Marvin Company
*W. B. Miller	Moore Products Company
*G. Platt	Bechtel Power Corporation
*J. R. Williams	Stearns Catalytic Corporation

*Director Emeritus

TABLE OF CONTENTS

Section	Title	Page
1.	Purpose	9
2.	Scope	9
2.1	General	9
2.2	Application to Industries	9
2.3	Application to Work Activities	9
2.4	Application to Classes of Instrumentation and to Instrument Functions	9
2.5	Extent of Functional Identification	9
2.6	Extent of Loop Identification	9
3.	Definitions	9
4.	Outline of the Identification System	12
4.1	General	12
4.2	Functional Identification	12
4.3	Loop Identification	13
4.4	Symbols	13
5.	Tables	14
5.1	Notes for Table 1	15
5.2	Table 1 - Identification Letters	17
5.3	Table 2 - Typical Letter Combinations	18
5.4	Table 3 - Function Blocks - Function Designations	19
6.	Drawings	23
6.1	Cautionary Notes	23
6.2	Instrument Line Symbols	24
6.3	General Instrument or Function Symbols	25
6.4	Control Valve Body Symbols, Damper Symbols	27
6.5	Actuator Symbols	28
6.6	Symbols for Self-Actuated Regulators, Valves, and Other Devices	30
6.7	Symbols for Actuator Action in the Event of Actuator Power Failure	33
6.8	Primary Element Symbols	34
6.9	Examples - Functions	44
6.10	Examples - Miscellaneous Combinations	52
6.11	Example - Complex Combinations	57
6.12	Example - Degree of Detail	58
	Index	61

LIST OF FIGURES

Figure	Title	Page
1.	Tag Numbers	12

1 PURPOSE

The purpose of this standard is to establish a uniform means of designating instruments and instrumentation systems used for measurement and control. To this end, a designation system that includes symbols and an identification code is presented.

2 SCOPE

2.1 General

2.1.1 The procedural needs of various users are different. The standard recognizes these needs, when they are consistent with the objectives of the standard, by providing alternative symbolism methods. A number of examples are provided for adding information or simplifying the symbolism, as desired.

2.1.2 Process equipment symbols are not part of this standard, but are included only to illustrate applications of instrumentation symbols.

2.2 Application to Industries

2.2.1 The standard is suitable for use in the chemical, petroleum, power generation, air conditioning, metal refining, and numerous other, process industries.

2.2.2 Certain fields, such as astronomy, navigation, and medicine, use very specialized instruments that are different from the conventional industrial process instruments. No specific effort was made to have the standard meet the requirements of those fields. However, it is expected that the standard will be flexible enough to meet many of the needs of special fields.

2.3 Application to Work Activities

2.3.1 The standard is suitable for use whenever any reference to an instrument or to a control system function is required for the purposes of symbolization and identification. Such references may be required for the following uses, as well as others:

- Design sketches
- Teaching examples
- Technical papers, literature, and discussions
- Instrumentation system diagrams, loop diagrams, logic diagrams
- Functional descriptions
- Flow diagrams: Process, Mechanical, Engineering, Systems, Piping (Process) and Instrumentation
- Construction drawings
- Specifications, purchase orders, manifests, and other lists
- Identification (tagging) of instruments and control functions
- Installation, operating and maintenance instructions, drawings, and records

2.3.2 The standard is intended to provide sufficient information to enable anyone reviewing any document depicting process measurement and control (who has a reasonable amount of process knowledge) to understand the means of measurement and control of the process. The detailed knowledge of a specialist in instrumentation is not a prerequisite to this understanding.

2.4 Application to Classes of Instrumentation and to Instrument Functions

The symbolism and identification methods provided in this standard are applicable to all classes of process measurement and control instrumentation. They can be used not only to describe discrete instruments and their functions, but also to describe the analogous functions of systems which are variously termed "shared display," "shared control," "distributed control," and "computer control."

2.5 Extent of Functional Identification

The standard provides for the identification and symbolization of the key functions of an instrument. Additional details of the instrument are better described in a suitable specification, data sheet, or other document intended for those requiring such details.

2.6 Extent of Loop Identification

The standard covers the identification of an instrument and all other instruments or control functions associated with it in a loop. The user is free to apply additional identification—by serial number, unit number, area number, plant number, or by other means.

3 DEFINITIONS

For the purpose of understanding this standard, the following definitions apply. For a more complete treatment, see ISA-S51.1 and the ISA-S75 series of standards. Terms italicized in a definition are also defined in this section.

Accessible - A term applied to a device or *function* that can be used or be seen by an operator for the purpose of performing control actions, *e.g., set point* changes, auto-manual transfer, or on-off actions.

Alarm - A device or *function* that signals the existence of an abnormal condition by means of an audible or visible discrete change, or both, intended to attract attention.

It is not recommended that the term *alarm switch* or *alarm* be used to designate a device whose operation is simply to close or open a circuit that may or may not be used for normal or abnormal interlock, start-up, shutdown, actuation of a *pilot light* or an *alarm* device, or the like. The first device is properly designated as a level *switch,* a flow *switch, etc.,* because "switching" is what the device does. The device may be designated as an *alarm* only if the device itself contains the *alarm function*. [*See also* Table 1, note (13).]

Assignable - A term applied to a feature permitting the channeling (or directing) of a signal from one device to another without the need for switching, patching, or changes in wiring.

Auto-Manual Station - Synonym for *control station*.

Balloon - Synonym for *bubble*.

Behind the Panel - A term applied to a location that is within an area that contains (1) the *instrument panel*, (2) its associated rack-mounted hardware, or (3) is enclosed within the *panel*. *Behind the panel* devices are not *accessible* for the operator's normal use, and are not designated as *local* or front-of-*panel-mounted*. In a very broad sense, "*behind the panel*" is equivalent to "not normally *accessible* to the operator."

Binary - A term applied to a signal or device that has only two discrete positions or states. When used in its simplest form, as in "*binary* signal" (as opposed to "analog signal"), the term denotes an "on-off" or "high-low" state, *i.e.*, one which does not represent continuously varying quantities.

Board - Synonym for *panel*.

Bubble - The circular symbol used to denote and identify the purpose of an *instrument* or *function*. It may contain a tag number. Synonym for *balloon*.

Computing Device - A device or *function* that performs one or more calculations or logic operations, or both, and transmits one or more resultant output signals. A *computing device* is sometimes called a computing *relay*.

Configurable - A term applied to a device or system whose functional characteristics can be selected or rearranged through programming or other methods. The concept excludes rewiring as a means of altering the configuration.

Controller - A device having an output that varies to regulate a controlled variable in a specified manner. A *controller* may be a self-contained analog or *digital instrument*, or it may be the equivalent of such an *instrument* in a shared-control system.

An automatic *controller* varies its output automatically in response to a direct or indirect input of a measured *process variable*. A manual *controller* is a *manual loading station*, and its output is not dependent on a measured *process variable* but can be varied only by manual adjustment.

A *controller* may be integral with other functional elements of a control *loop*.

Control Station - A *manual loading station* that also provides switching between manual and automatic control modes of a control *loop*. It is also known as an *auto-manual station*. In addition, the operator interface of a *distributed control system* may be regarded as a *control station*.

Control Valve - A device, other than a common, hand-actuated ON-OFF valve or self-actuated check valve, that directly manipulates the flow of one or more fluid process streams.

It is expected that use of the designation "hand *control valve*" will be limited to hand-actuated valves that (1) are used for process throttling, or (2) require *identification* as an *instrument*.

Converter - A device that receives information in one form of an instrument signal and transmits an output signal in another form.

An *instrument* which changes a sensor's output to a standard signal is properly designated as a *transmitter*, not a *converter*. Typically, a temperature element (*TE*) may connect to a *transmitter* (*TT*), not to a *converter* (*TY*).

A converter is also referred to as a *transducer;* however, "*transducer*" is a completely general term, and its use specifically for signal conversion is not recommended.

Digital - A term applied to a signal or device that uses *binary* digits to represent continuous values or discrete states.

Distributed Control System - A system which, while being functionally integrated, consists of subsystems which may be physically separate and remotely located from one another.

Final Control Element - The device that directly controls the value of the manipulated variable of a control *loop*. Often the *final control element* is a *control valve*.

Function - The purpose of, or an action performed by, a device.

Identification - The sequence of letters or digits, or both, used to designate an individual *instrument* or *loop*.

Instrument - A device used directly or indirectly to measure and/or control a variable. The term includes *primary elements*, *final control elements*, *computing devices*, and electrical devices such as annunciators, *switches*, and pushbuttons. The term does not apply to parts (*e.g.*, a receiver bellows or a resistor) that are internal components of an *instrument*.

Instrumentation - A collection of *instruments* or their application for the purpose of observation, *measurement*, control, or any combination of these.

Local - The location of an *instrument* that is neither in nor on a *panel* or console, nor is it mounted in a control room. *Local instruments* are commonly in the vicinity of a *primary element* or a *final control element*. The word "field" is often used synonymously with *local*.

Local Panel - A *panel* that is not a central or main *panel*. *Local panels* are commonly in the vicinity of plant subsystems or sub-areas. The term *"local panel instrument"* should not be confused with *"local instrument."*

Loop - A combination of two or more *instruments* or control *functions* arranged so that signals pass from one to another for the purpose of *measurement* and/or control of a *process variable*.

Manual Loading Station - A device or *function* having a manually adjustable output that is used to actuate one or more remote devices. The station does not provide switching between manual and automatic control modes of a control *loop (see controller* and *control station*). The station may have integral indicators, lights, or other features. It is also known as a manual station or a manual loader.

Measurement - The determination of the existence or the magnitude of a variable.

Monitor - A general term for an *instrument* or *instrument* system used to measure or sense the status or magnitude of one or more variables for the purpose of deriving useful information. The term *monitor* is very unspecific—sometimes meaning analyzer, indicator, or *alarm*. *Monitor* can also be used as a verb.

Monitor Light - Synonym for *pilot light*.

Panel - A structure that has a group of *instruments* mounted on it, houses the operator-process interface, and is chosen to have a unique designation. The *panel* may consist of one or more sections, cubicles, consoles, or desks. Synonym for *board*.

Panel-Mounted - A term applied to an *instrument* that is mounted on a *panel* or console and is *accessible* for an operator's normal use. A *function* that is normally *accessible* to an operator in a *shared-display* system is the equivalent of a discrete *panel-mounted* device.

Pilot Light - A light that indicates which of a number of normal conditions of a system or device exists. It is unlike an *alarm* light, which indicates an abnormal condition. The *pilot light* is also known as a *monitor light*.

Primary Element - Synonym for *sensor*.

Process - Any operation or sequence of operations involving a change of energy, state, composition, dimension, or other properties that may be defined with respect to a datum.

Process Variable - Any variable property of a *process*. The term *process variable* is used in this standard to apply to all variables other than *instrument* signals.

Program - A repeatable sequence of actions that defines the status of outputs as a fixed relationship to a set of inputs.

Programmable Logic Controller - A *controller,* usually with multiple inputs and outputs, that contains an alterable *program*.

Relay - A device whose *function* is to pass on information in an unchanged form or in some modified form. *Relay* is often used to mean *computing device*. The latter term is preferred.

The term *"relay"* also is applied specifically to an electric, pneumatic, or hydraulic *switch* that is actuated by a signal. The term also is applied to *functions* performed by a *relay*.

Scan - To sample, in a predetermined manner, each of a number of variables intermittently. The *function* of a scanning device is often to ascertain the state or value of a variable. The device may be associated with other *functions* such as recording and alarming.

Sensor - That part of a *loop* or *instrument* that first senses the value of a process variable, and that assumes a corresponding, predetermined, and intelligible state or output. The *sensor* may be separate from or integral with another functional element of a *loop*. The *sensor* is also known as a detector or *primary element*.

Set Point - An input variable that sets the desired value of the controlled variable. The *set point* may be manually set, automatically set, or programmed. Its value is expressed in the same units as the controlled variable.

Shared Controller - A *controller,* containing preprogrammed algorithms that are usually *accessible, configurable,* and *assignable*. It permits a number of *process variables* to be controlled by a single device.

Shared Display - The operator interface device (usually a video screen) used to display *process* control information from a number of sources at the command of the operator.

Switch - A device that connects, disconnects, selects, or transfers one or more circuits and is not designated as a *controller*, a *relay*, or a *control valve*. As a verb, the term is also applied to the *functions* performed by *switches*.

Test Point - A *process* connection to which no *instrument* is permanently connected, but which is intended for the temporary or intermittent connection of an *instrument*.

Transducer - A general term for a device that receives information in the form of one or more physical quantities, modifies the information and/or its form, if required, and produces a resultant output signal. Depending on the application, the *transducer* can be a *primary element, transmitter, relay, converter* or other device. Because the term *"transducer"* is not specific, its use for specific applications is not recommended.

Transmitter - A device that senses a *process variable* through the medium of a sensor and has an output whose steady-state value varies only as a predetermined *function* of the *process variable*. The *sensor* may or may not be integral with the *transmitter*.

4 OUTLINE OF THE IDENTIFICATION SYSTEM

4.1 General

4.1.1 Each instrument or function to be identified is designated by an alphanumeric code or tag number as shown in Figure 1. The loop identification part of the tag number generally is common to all instruments or functions of the loop. A suffix or prefix may be added to complete the identification. Typical identification is shown in Figure 1.

TYPICAL TAG NUMBER

TIC 103	-	Instrument Identification or Tag Number
T 103	-	Loop Identification
103	-	Loop Number
TIC	-	Functional Identification
T	-	First-letter
IC	-	Succeeding-Letters

EXPANDED TAG NUMBER

10-PAH-5A	-	Tag Number
10	-	Optional Prefix
A	-	Optional Suffix

Note: Hyphens are optional as separators.

Figure 1. Tag Numbers

4.1.2 The instrument loop number may include coded information, such as plant area designation. It is also possible to set aside specific series of numbers to designate special functions; for instance, the series 900 to 999 could be used for loops whose primary function is safety-related.

4.1.3 Each instrument may be represented on diagrams by a symbol. The symbol may be accompanied by a tag number.

4.2 Functional Identification

4.2.1 The functional identification of an instrument or its functional equivalent consists of letters from Table 1 and includes one first-letter (designating the measured or initiating variable) and one or more succeeding-letters (identifying the functions performed).

4.2.2 The functional identification of an instrument is made according to the function and not according to the construction. Thus, a differential-pressure recorder used for flow measurement is identified by *FR;* a pressure indicator and a pressure-actuated switch connected to the output of a pneumatic level transmitter are identified by *LI* and *LS*, respectively.

4.2.3 In an instrument loop, the first-letter of the functional identification is selected according to the measured or initiating variable, and not according to the manipulated variable. Thus, a control valve varying flow according to the dictates of a level controller is an *LV*, not an *FV*.

4.2.4 The succeeding-letters of the functional identification designate one or more readout or passive functions and/or output functions. A modifying-letter may be used, if required, in addition to one or more other succeeding-letters. Modifying-letters may modify either a first-letter or succeeding-letters, as applicable. Thus, *TDAL* contains two modifiers. The letter *D* changes the measured variable *T* into a new variable, "differential temperature." The letter *L* restricts the readout function *A*, alarm, to represent a low alarm only.

4.2.5 The sequence of identification letters begins with a first-letter selected according to Table 1. Readout or passive functional letters follow in any order, and output functional letters follow these in any sequence, except that output letter *C* (control) precedes output letter *V* (valve), *e.g.*, *PCV*, a self-actuated control valve. However, modifying-letters, if used, are interposed so that they are placed immediately following the letters they modify.

4.2.6 A multiple function device may be symbolized on a diagram by as many bubbles as there are measured variables, outputs, and/or functions. Thus, a temperature controller with a switch may be identified by two tangent bubbles—one inscribed *TIC-3* and one inscribed *TSH-3*. The instrument would be designated *TIC/TSH-3* for all uses in writing or reference. If desired, however, the abbreviation *TIC-3* may serve for general identification or for purchasing, while *TSH-3* may be used for electric circuit diagrams.

4.2.7 The number of functional letters grouped for one instrument should be kept to a minimum according to the judgment of the user. The total number of letters within one group should not exceed four. The number within a group may be kept to a minimum by:

(1) Arranging the functional letters into subgroups. This practice is described in Section 4.2.6 for instruments having more than one measured variable or input, but it may also be used for other instruments.

(2) Omitting the *I* (indicate) if an instrument both indicates and records the same measured variable.

4.2.8 All letters of the functional identification are uppercase.

4.3 Loop Identification

4.3.1 The loop identification consists of a first-letter and a number. Each instrument within a loop has assigned to it the same loop number and, in the case of parallel numbering, the same first-letter. Each instrument loop has a unique loop identification. An instrument common to two or more loops should carry the identification of the loop which is considered predominant.

4.3.1 Loop numbering may be parallel or serial. Parallel numbering involves starting a numerical sequence for each new first-letter, *e.g., TIC-100, FRC-100, LIC-100, AI-100, etc.* Serial numbering involves using a single sequence of numbers for a project or for large sections of a project, regardless of the first-letter of the loop identification, *e.g., TIC-100, FRC-101, LIC-102, AI-103, etc.* A loop numbering sequence may begin with *1* or any other convenient number, such as *001*, *301* or *1201*. The number may incorporate coded information; however, simplicity is recommended.

4.3.3 If a given loop has more than one instrument with the same functional identification, a suffix may be appended to the loop number, *e.g., FV-2A, FV-2B, FV-2C, etc.*, or *TE-25-1, TE-25-2, etc.* However, it may be more convenient or logical in a given instance to designate a pair of flow transmitters, for example, as *FT-2* and *FT-3* instead of *FT-2A* and *FT-2B*. The suffixes may be applied according to the following guidelines:

(1) An uppercase suffix letter should be used, *i.e., A, B, C, etc.*

(2) For an instrument such as a multipoint temperature recorder that prints numbers for point identification, the primary elements may be numbered *TE-25-1, TE-25-2, TE-25-3, etc.*, corresponding to the point identification number.

(3) Further subdivisions of a loop may be designated by serially alternating suffix letters and numbers. (*See* Section 6.9R(3).)

4.3.4 An instrument that performs two or more functions may be designated by all of its functions. For example, a flow recorder *FR-2* with a pressure pen *PR-4* may be designated *FR-2/PR-4*. A two-pen pressure recorder may be *PR-7/8*, and a common annunciator window for high and low temperature alarms may be *TAHL-21*. Note that the slash is not necessary when distinctly separate devices are not present.

4.3.5 Instrument accessories such as purge meters, air sets, and seal pots that are not explicitly shown on a diagram but that need a designation for other purposes should be tagged individually according to their functions and should use the same loop identification as the instrument they directly serve. Application of such a designation does not imply that the accessory must be shown on the diagram. Alternatively, the accessories may use the identical tag number as that of their associated instrument, but with clarifying words added. Thus an orifice flange union associated with orifice plate *FE-7* should be tagged *FX-7*, but may be designated *FE-7 FLANGES*. A purge meter associated with pressure gauge *PI-8* may be tagged *PI-8 PURGE*. A thermowell used with thermometer *TI-9* should be tagged *TW-9*, but may be tagged *TI-9 THERMOWELL*.

The rules for loop identification need not be applied to instruments and accessories that are purchased in bulk quantities if it is the user's practice to identify these items by other means.

4.4 Symbols

4.4.1 The examples in this standard illustrate the symbols that are intended to depict instrumentation on diagrams and drawings. Methods of symbolization and identification are demonstrated. The examples show identification that is typical for the pictured instrument or functional interrelationships. The symbols indicating the various instruments or functions have been applied in typical ways in the illustrations. This usage does not imply, however, that the applications or designations of the instruments or functions are restricted in any way. No inference should be drawn that the choice of any of the schemes for illustration constitutes a recommendation for the illustrated methods of measurement or control. Where alternative symbols are shown without a statement of preference, the relative sequence of symbols does not imply a preference.

4.4.2 The bubble may be used to tag distinctive symbols, such as those for control valves, when such tagging is desired. In such instances, the line connecting the bubble to the instrument symbol is drawn close to, but not touching, the symbol. In other instances, the bubble serves to represent the instrument proper.

4.4.3 A distinctive symbol whose relationship to the remainder of the loop is easily apparent from a diagram need not be individually tagged on the diagram. For example, an orifice flange or a control valve that is part of a larger system need not be shown with a tag number on a diagram. Also, where there is a primary element connected to another instrument on a diagram, use of a symbol to represent the primary element on the diagram is optional.

4.4.4 A brief explanatory notation may be added adjacent to a symbol or line to clarify the function of an item. For instance, the notations *3-9 psig* and *9-15 psig* adjacent to the signal lines to two valves operating in split range, taken together with the symbols for the failure modes, allow complete understanding of the intent. Similarly, when two valves are operated in a diverting or mixing mode from a common signal, the notations *3-15 psig* and *15-3 psig*, together with the failure modes, allow understanding of the function.

4.4.5 The sizes of the tagging bubbles and the miscellaneous symbols shown in the examples are the sizes generally recommended; however, the optimum sizes may vary depending on whether or not the finished diagram is to be reduced in size and depending on the number of characters that are expected in the instrument tagging designation. The sizes of the other symbols may be selected as appropriate to accompany the symbols of other equipment on a diagram.

4.4.6 Aside from the general drafting requirements for neatness and legibility, symbols may be drawn with any orientation. Likewise, signal lines may be drawn on a diagram entering or leaving the appropriate part of a symbol at any angle. However, the function block designators of Table 3 and the tag numbers should always be drawn with a horizontal orientation. Directional arrowheads should be added to signal lines when needed to clarify the direction of flow of information. The judicious use of such arrowheads, especially on complex drawings, will often facilitate understanding of the system.

4.4.7 The electrical, pneumatic, or other power supply to an instrument is not expected to be shown unless it is essential to an understanding of the operation of the instrument or the loop.

4.4.8 In general, one signal line will suffice to represent the interconnections between two instruments on flow diagrams even though they may be connected physically by more than one line.

4.4.9 The sequence in which the instruments or functions of a loop are connected on a diagram should reflect the functional logic or information flow, although this arrangement will not necessarily correspond to the signal connection sequence. Thus, an electronic loop using analog voltage signals requires parallel wiring, while a loop using analog current signals requires series interconnections. However, the diagram in both instances should be drawn as though all the wiring were parallel, to show the functional interrelationships clearly while keeping the presentation independent of the type of instrumentation finally installed. The correct interconnections are expected to be shown on a suitable diagram.

4.4.10 The degree of detail to be applied to each document or sketch is entirely at the discretion of the user of the standard. The symbols and designations in this standard can depict both hardware and function. Sketches and technical papers will usually contain highly simplified symbolism and identification. Process flow diagrams will usually be less detailed than engineering flow diagrams. Engineering flow diagrams may show all in-line components, but may differ from user to user in the amount of off-line detail shown. In any case, consistency should be established for each application. The terms *simplified, conceptual,* and *detailed* as applied to the diagrams of 6.12 were chosen to represent a cross section of typical usage. Each user must establish the degree of detail that fulfills the purposes of the specific document or sketch being generated.

4.4.11 It is common practice for engineering flow diagrams to omit the symbols of interlock-hardware components that are actually necessary for a working system, particularly when symbolizing electric interlock systems. For example, a level switch may be shown as tripping a pump, or separate flow and pressure switches may be shown as actuating a solenoid valve or other interlock devices. In both instances, auxiliary electrical relays and other components may be considered details to be shown elsewhere. By the same token, a current transformer sometimes will be omitted and its receiver shown connected directly to the process—in this case the electric motor.

4.4.12 Because the distinctions between shared display/shared control and computer functions are sometimes blurred, in choosing symbols to represent them the user must rely on manufacturers' definitions, usage in a particular industry, and personal judgment.

5 TABLES

The purpose of Section 5, Tables, is to define certain of the building blocks of the identification and symbolic representation system used in this standard in a concise, easily-referenced manner.

Table 1, Identification Letters, together with the Notes for Table 1, define and explain the individual letter designators used as functional identifiers in accordance with the rules of Section 4.2, Functional Identification.

Table 2, Typical Letter Combinations, attempts to facilitate the task of choosing acceptable combinations of identifying letters.

Table 3, Function Blocks - Function Designations, is an adaptation of the SAMA (Scientific Apparatus Manufacturers Association) method of functional diagramming. Two basic uses are found for these symbols: as standalone function blocks on conceptual diagrams, or as flags which designate functions performed by bubbles on more detailed drawings. A third use is a combination of the first two and is found in shared control systems where, for instance, the measured variable signal line enters a square root function block that is drawn adjacent to a shared controller.

Two omissions will be noted: The SAMA symbol for *Transfer* and that for an *Analog Signal Generator*. Since the ultimate use of ISA-S5.1 symbolism usually requires identification to be associated with a symbol, it is advisable to use the *HIC* (manual loader) bubble for an analog signal generator and an *HS* (hand switch) with or without a relay bubble for a transfer function.

5.1 Notes for Table 1

(1) A "user's choice" letter is intended to cover unlisted meanings that will be used repetitively in a particular project. If used, the letter may have one meaning as a first-letter and another meaning as a succeeding-letter. The meanings need to be defined only once in a legend, or other place, for that project. For example, the letter *N* may be defined as "modulus of elasticity" as a first-letter and "oscilloscope" as a succeeding-letter.

(2) The unclassified letter *X* is intended to cover unlisted meanings that will be used only once or used to a limited extent. If used, the letter may have any number of meanings as a first-letter and any number of meanings as a succeeding-letter. Except for its use with distinctive symbols, it is expected that the meanings will be defined outside a tagging bubble on a flow diagram. For example, *XR-2* may be a stress recorder and *XX-4* may be a stress oscilloscope.

(3) The grammatical form of the succeeding-letter meanings may be modified as required. For example, "indicate" may be applied as "indicator" or "indicating," "transmit" as "transmitter" or "transmitting," *etc.*

(4) Any first-letter, if used in combination with modifying letters *D* (differential), *F* (ratio), *M* (momentary), *K* (time rate of change), *Q* (integrate or totalize), or any combination of these is intended to represent a new and separate measured variable, and the combination is treated as a first-letter entity. Thus, instruments *TDI* and *TI* indicate two different variables, namely, differential-temperature and temperature. Modifying letters are used when applicable.

(5) First-letter *A* (analysis) covers all analyses not described by a "user's choice" letter. It is expected that the type of analysis will be defined outside a tagging bubble.

(6) Use of first-letter *U* for "multivariable" in lieu of a combination of first-letters is optional. It is recommended that nonspecific variable designators such as *U* be used sparingly.

(7) The use of modifying terms "high," "low," "middle" or "intermediate," and "scan" is optional.

(8) The term "safety" applies to emergency protective primary elements and emergency protective final control elements only. Thus, a self-actuated valve that prevents operation of a fluid system at a higher-than-desired pressure by bleeding fluid from the system is a back-pressure-type *PCV*, even if the valve is not intended to be used normally. However, this valve is designated as a *PSV* if it is intended to protect against emergency conditions, *i.e.*, conditions that are hazardous to personnel and/or equipment and that are not expected to arise normally.

The designation *PSV* applies to all valves intended to protect against emergency pressure conditions regardless of whether the valve construction and mode of operation place them in the category of the safety valve, relief valve, or safety relief valve. A rupture disc is designated *PSE*.

(9) The passive function *G* applies to instruments or devices that provide an uncalibrated view, such as sight glasses and television monitors.

(10) "Indicate" normally applies to the readout—analog or digital—of an actual measurement. In the case of a manual loader, it may be used for the dial or setting indication, *i.e.*, for the value of the initiating variable.

(11) A pilot light that is part of an instrument loop should be designated by a first-letter followed by the succeeding-letter *L*. For example, a pilot light that indicates an expired time period should be tagged *KQL*. If it is desired to tag a pilot light that is not part of an instrument loop, the light is designated in the same way. For example, a running light for an electric motor may be tagged *EL*, assuming voltage to be the appropriate measured variable, or *YL*, assuming the operating status is being monitored. The unclassified variable *X* should be used only for applications which are limited in extent. The designation *XL* should not be used for motor running lights, as these are commonly numerous. It is permissible to use the user's choice letters *M, N* or *O* for a motor running light when the meaning is previously defined. If *M* is used, it must be clear that the letter does not stand for the word "motor," but for a monitored state.

(12) Use of a succeeding-letter *U* for "multifunction" instead of a combination of other functional letters is optional. This nonspecific function designator should be used sparingly.

(13) A device that connects, disconnects, or transfers one or more circuits may be either a switch, a relay, an ON-OFF controller, or a control valve, depending on the application.

If the device manipulates a fluid process stream and is not a hand-actuated ON-OFF block valve, it is designated as a control valve. It is incorrect to use the succeeding-letters *CV* for anything other than a self-actuated control valve. For all applications other than fluid process streams, the device is designated as follows:

> A switch, if it is actuated by hand.
>
> A switch or an ON-OFF controller, if it is automatic and is the first such device in a loop. The term "switch" is generally used if the device is used for alarm, pilot light, selection, interlock, or safety.

The term "controller" is generally used if the device is used for normal operating control.

A relay, if it is automatic and is not the first such device in a loop, *i.e.*, it is actuated by a switch or an ON-OFF controller.

(14) It is expected that the functions associated with the use of succeeding-letter *Y* will be defined outside a bubble on a diagram when further definition is considered necessary. This definition need not be made when the function is self-evident, as for a solenoid valve in a fluid signal line.

(15) The modifying terms "high," and "low," and "middle" or "intermediate" correspond to values of the measured variable, not to values of the signal, unless otherwise noted. For example, a high-level alarm derived from a reverse-acting level transmitter signal should be an *LAH*, even though the alarm is actuated when the signal falls to a low value. The terms may be used in combinations as appropriate. (*See* Section 6.9A.)

(16) The terms "high" and "low," when applied to positions of valves and other open-close devices, are defined as follows: "high" denotes that the valve is in or approaching the fully open position, and "low" denotes that it is in or approaching the fully closed position.

(17) The word "record" applies to any form of permanent storage of information that permits retrieval by any means.

(18) For use of the term "transmitter" versus "converter," see the definitions in Section 3.

(19) First-letter *V*, "vibration or mechanical analysis," is intended to perform the duties in machinery monitoring that the letter *A* performs in more general analyses. Except for vibration, it is expected that the variable of interest will be defined outside the tagging bubble.

(20) First-letter *Y* is intended for use when control or monitoring responses are event-driven as opposed to time- or time schedule-driven. The letter *Y*, in this position, can also signify presence or state.

(21) Modifying-letter *K*, in combination with a first-letter such as *L*, *T*, or *W*, signifies a time rate of change of the measured or initiating variable. The variable *WKIC*, for instance, may represent a rate-of-weight-loss controller.

(22) Succeeding-letter *K* is a user's option for designating a control station, while the succeeding-letter *C* is used for describing automatic or manual controllers. (*See* Section 3, Definitions.)

TABLE 1
IDENTIFICATION LETTERS

	FIRST-LETTER (4)		SUCCEEDING-LETTERS (3)		
	MEASURED OR INITIATING VARIABLE	MODIFIER	READOUT OR PASSIVE FUNCTION	OUTPUT FUNCTION	MODIFIER
A	Analysis(5,19)		Alarm		
B	Burner, Combustion		User's Choice(1)	User's Choice(1)	User's Choice(1)
C	User's Choice(1)			Control(13)	
D	User's Choice(1)	Differential(4)			
E	Voltage		Sensor (Primary Element)		
F	Flow Rate	Ratio (Fraction)(4)			
G	User's Choice(1)		Glass, Viewing Device(9)		
H	Hand				High(7,15,16)
I	Current (Electrical)		Indicate(10)		
J	Power	Scan(7)			
K	Time, Time Schedule	Time Rate of Change(4,21)		Control Station (22)	
L	Level		Light(11)		Low(7,15,16)
M	User's Choice(1)	Momentary(4)			Middle, Intermediate(7,15)
N	User's Choice(1)		User's Choice(1)	User's Choice(1)	User's Choice(1)
O	User's Choice(1)		Orifice, Restriction		
P	Pressure, Vacuum		Point (Test) Connection		
Q	Quantity	Integrate, Totalize(4)			
R	Radiation		Record(17)		
S	Speed, Frequency	Safety(8)		Switch(13)	
T	Temperature			Transmit(18)	
U	Multivariable(6)		Multifunction(12)	Multifunction(12)	Multifunction(12)
V	Vibration, Mechanical Analysis(19)			Valve, Damper, Louver(13)	
W	Weight, Force		Well		
X	Unclassified(2)	X Axis	Unclassified(2)	Unclassified(2)	Unclassified(2)
Y	Event, State or Presence(20)	Y Axis		Relay, Compute, Convert(13,14,18)	
Z	Position, Dimension	Z Axis		Driver, Actuator, Unclassified Final Control Element	

Note: Numbers in parentheses refer to specific explanatory notes on pages 15 and 16.

TABLE 2
TYPICAL LETTER COMBINATIONS

First-Letters	Initiating or Measured Variable	Controllers				Readout Devices		Switches and Alarm Devices*			Transmitters			Solenoids, Relays, Computing Devices	Primary Element	Test Point	Well or Probe	Viewing Device, Glass	Safety Device	Final Element
		Recording	Indicating	Blind	Self-Actuated Control Valves	Recording	Indicating	High**	Low	Comb	Recording	Indicating	Blind							
A	Analysis	ARC	AIC	AC		AR	AI	ASH	ASL	ASHL	ART	AIT	AT	AY	AE	AP	AW			AV
B	Burner/ Combustion	BRC	BIC	BC		BR	BI	BSH	BSL	BSHL	BRT	BIT	BT	BY	BE		BW	BG		BZ
C	User's Choice																			
D	User's Choice																			
E	Voltage	ERC	EIC	EC		ER	EI	ESH	ESL	ESHL	ERT	EIT	ET	EY	EE					EZ
F	Flow Rate	FRC	FIC	FC	FCV, FICV	FR	FI	FSH	FSL	FSHL	FRT	FIT	FT	FY	FE	FP		FG		FV
FQ	Flow Quantity	FQRC	FQIC			FQR	FQI	FQSH	FQSL			FQIT	FQT	FQY	FQE					FQV
FF	Flow Ratio	FFRC	FFIC	FFC		FFR	FFI	FFSH	FFSL						FE					FFV
G	User's Choice																			
H	Hand		HIC	HC						HS										HV
I	Current	IRC	IIC			IR	II	ISH	ISL	ISHL	IRT	IIT	IT	IY	IE					IZ
J	Power	JRC	JIC			JR	JI	JSH	JSL	JSHL	JRT	JIT	JT	JY	JE					JV
K	Time	KRC	KIC	KC	KCV	KR	KI	KSH	KSL	KSHL	KRT	KIT	KT	KY	KE					KV
L	Level	LRC	LIC	LC	LCV	LR	LI	LSH	LSL	LSHL	LRT	LIT	LT	LY	LE		LW	LG		LV
M	User's Choice																			
N	User's Choice																			
O	User's Choice																			
P	Pressure/ Vacuum	PRC	PIC	PC	PCV	PR	PI	PSH	PSL	PSHL	PRT	PIT	PT	PY	PE	PP			PSV, PSE	PV
PD	Pressure, Differential	PDRC	PDIC	PDC	PDCV	PDR	PDI	PDSH	PDSL		PDRT	PDIT	PDT	PDY	PE	PP				PDV
Q	Quantity	QRC	QIC			QR	QI	QSH	QSL	QSHL	QRT	QIT	QT	QY	QE					QZ
R	Radiation	RRC	RIC	RC		RR	RI	RSH	RSL	RSHL	RRT	RIT	RT	RY	RE		RW			RZ
S	Speed/ Frequency	SRC	SIC	SC	SCV	SR	SI	SSH	SSL	SSHL	SRT	SIT	ST	SY	SE					SV
T	Temperature	TRC	TIC	TC	TCV	TR	TI	TSH	TSL	TSHL	TRT	TIT	TT	TY	TE	TP	TW		TSE	TV
TD	Temperature, Differential	TDRC	TDIC	TDC	TDCV	TDR	TDI	TDSH	TDSL		TDRT	TDIT	TDT	TDY	TE	TP	TW			TDV
U	Multivariable					UR	UI							UY						UV
V	Vibration/ Machinery Analysis					VR	VI	VSH	VSL	VSHL	VRT	VIT	VT	VY	VE					VZ
W	Weight/Force	WRC	WIC	WC	WCV	WR	WI	WSH	WSL	WSHL	WRT	WIT	WT	WY	WE					WZ
WD	Weight/Force, Differential	WDRC	WDIC	WDC	WDCV	WDR	WDI	WDSH	WDSL		WDRT	WDIT	WDT	WDY	WE					WDZ
X	Unclassified																			
Y	Event/State/ Presence		YIC	YC		YR	YI	YSH	YSL				YT	YY	YE					YZ
Z	Position/ Dimension	ZRC	ZIC	ZC	ZCV	ZR	ZI	ZSH	ZSL	ZSHL	ZRT	ZIT	ZT	ZY	ZE					ZV
ZD	Gauging/ Deviation	ZDRC	ZDIC	ZDC	ZDCV	ZDR	ZDI	ZDSH	ZDSL		ZDRT	ZDIT	ZDT	ZDY	ZDE					ZDV

Note: This table is not all-inclusive.

*A, alarm, the annunciating device, may be used in the same fashion as S, switch, the actuating device.

**The letters H and L may be omitted in the undefined case.

Other Possible Combinations:

FO	(Restriction Orifice)	PFR	(Ratio)
FRK, HIK	(Control Stations)	KQI	(Running Time Indicator)
FX	(Accessories)	QQI	(Indicating Counter)
TJR	(Scanning Recorder)	WKIC	(Rate-of-Weight-Loss Controller)
LLH	(Pilot Light)	HMS	(Hand Momentary Switch)

5.4 S5.1

TABLE 3
FUNCTION BLOCKS-FUNCTION DESIGNATIONS

THE FUNCTION DESIGNATIONS ASSOCIATED WITH CONTROLLERS, COMPUTING DEVICES, CONVERTERS AND RELAYS MAY BE USED INDIVIDUALLY OR IN COMBINATION (ALSO, SEE TABLE 1, NOTE 14.). THE USE OF A BOX AVOIDS CONFUSION BY SETTING OFF THE SYMBOL FROM OTHER MARKINGS ON A DIAGRAM AND PERMITS THE FUNCTION TO BE USED AS A STAND-ALONE BLOCK ON CONCEPTUAL DESIGNS.

NO	FUNCTION	SYMBOL	MATH EQUATION	GRAPHIC REPRESENTATION	DEFINITION
1	SUMMING	Σ	$M = X_1 + X_2 + \ldots\ldots + X_n$	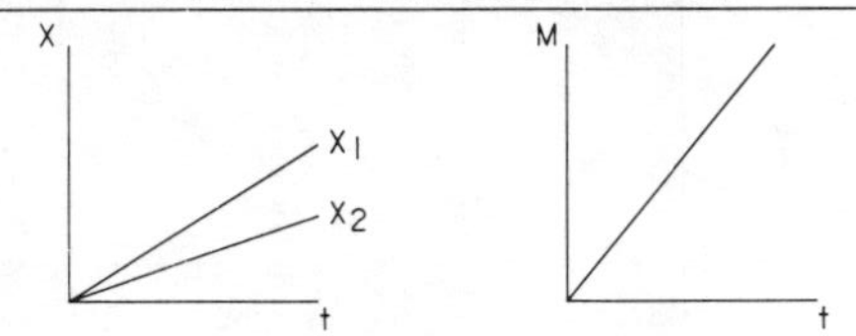	THE OUTPUT EQUALS THE ALGEBRAIC SUM OF THE INPUTS. (THE INPUTS MAY BE LABELED WITH POSITIVE OR NEGATIVE SIGNS).
2	AVERAGING	Σ/n	$M = \dfrac{X_1 + X_2 + \ldots\ldots + X_n}{n}$	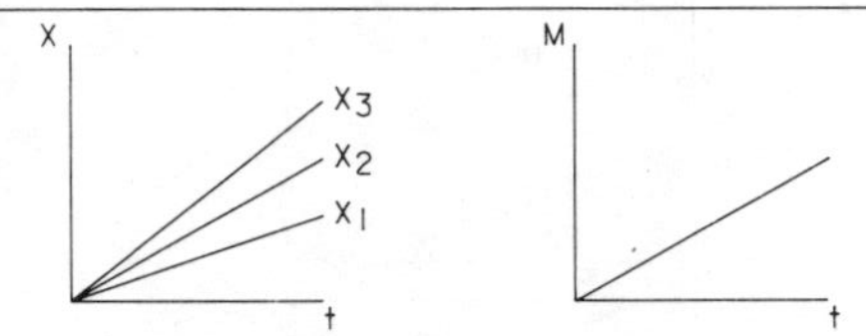	THE OUTPUT EQUALS THE ALGEBRAIC SUM OF THE INPUTS DIVIDED BY THE NUMBER OF INPUTS.
3	DIFFERENCE	Δ	$M = X_1 - X_2$	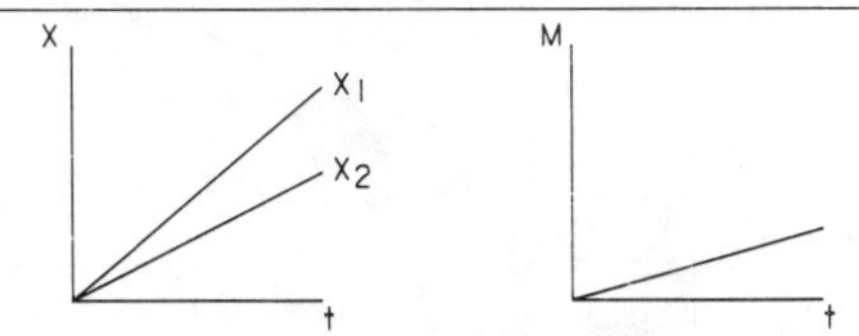	THE OUTPUT EQUALS THE ALGEBRAIC DIFFERENCE OF THE TWO INPUTS.
4	PROPORTIONAL	K 1:1 2:1	$M = KX$	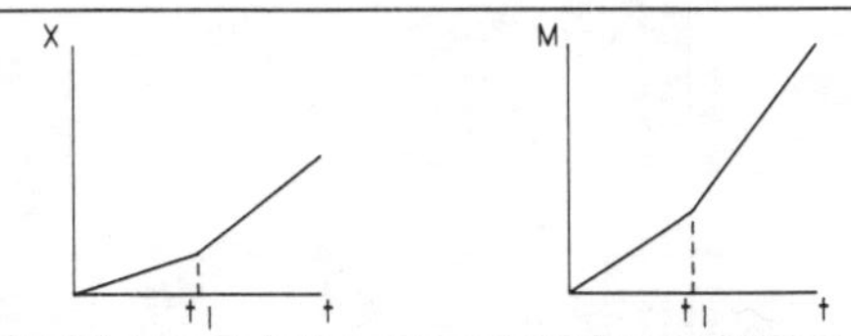	THE OUTPUT IS DIRECTLY PROPORTIONAL TO THE INPUT. IN THE CASE OF A VOLUME BOOSTER, "K" MAY BE REPLACED BY 1:1. FOR INTEGER GAINS, 2:1, 3:1, ETC., MAY BE SUBSTITUTED FOR K.
5	INTEGRAL	$\int$	$M = \frac{1}{T_I} \int X dt$	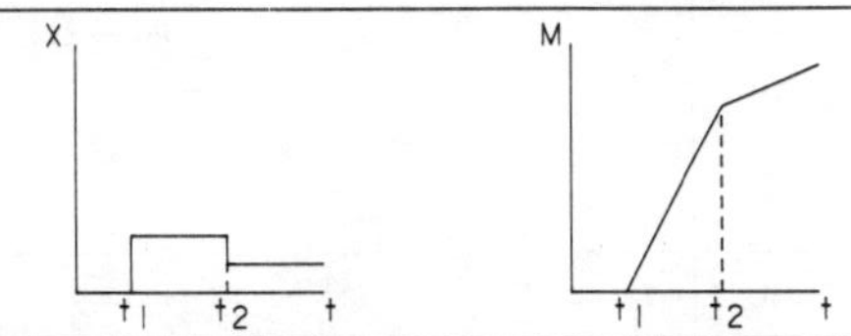	THE OUTPUT VARIES IN ACCORDANCE WITH BOTH MAGNITUDE AND DURATION OF THE INPUT. THE OUTPUT IS PROPORTIONAL TO THE TIME INTEGRAL OF THE INPUT.
6	DERIVATIVE	d/dt	$M = T_D \frac{dX}{dt}$	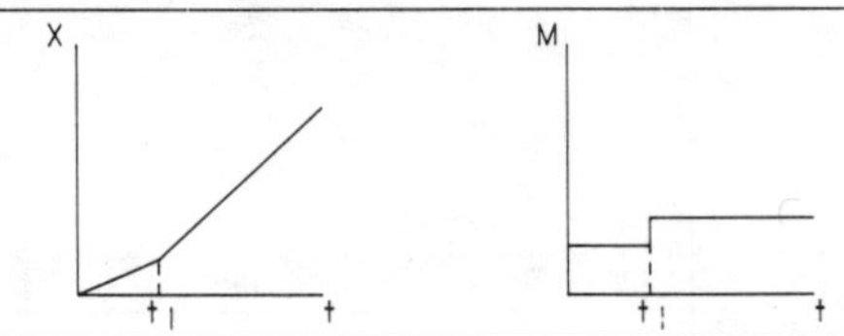	THE OUTPUT IS PROPORTIONAL TO THE RATE OF CHANGE (DERIVATIVE) OF THE INPUT.

5.4 TABLE 3 - CONTINUED

NO	FUNCTION	SYMBOL	MATH EQUATION	GRAPHIC REPRESENTATION	DEFINITION
7	MULTIPLYING	X	$M = X_1 X_2$	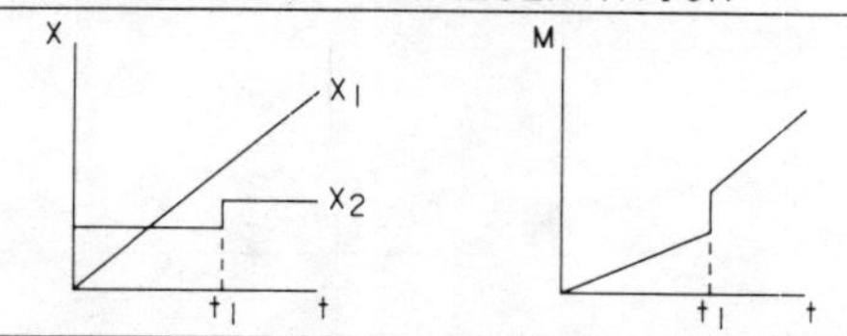	THE OUTPUT EQUALS THE PRODUCT OF THE TWO INPUTS.
8	DIVIDING	÷	$M = \frac{X_1}{X_2}$	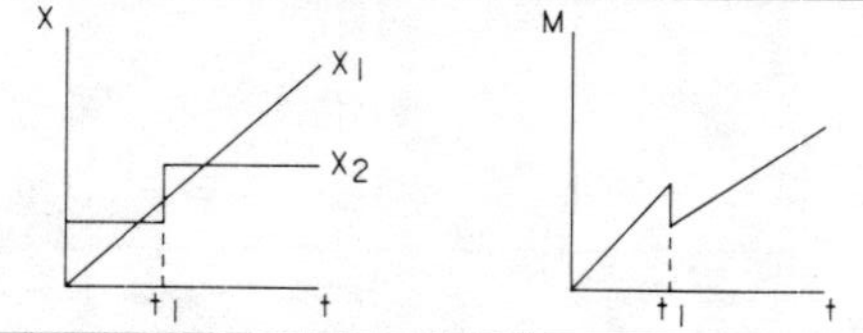	THE OUTPUT EQUALS THE QUOTIENT OF THE TWO INPUTS.
9	ROOT EXTRACTION	$\sqrt[n]{\;}$	$M = \sqrt[n]{X}$	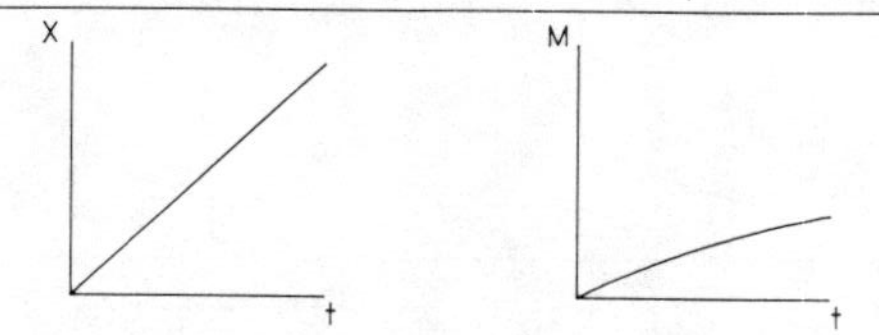	THE OUTPUT EQUALS THE ROOT (I.E., CUBE ROOT, FOURTH ROOT, 3/2 ROOT, ETC.) OF THE INPUT. IF n IS OMITTED, A SQUARE ROOT IS ASSUMED.
10	EXPONENTIAL	x^n	$M = X^n$	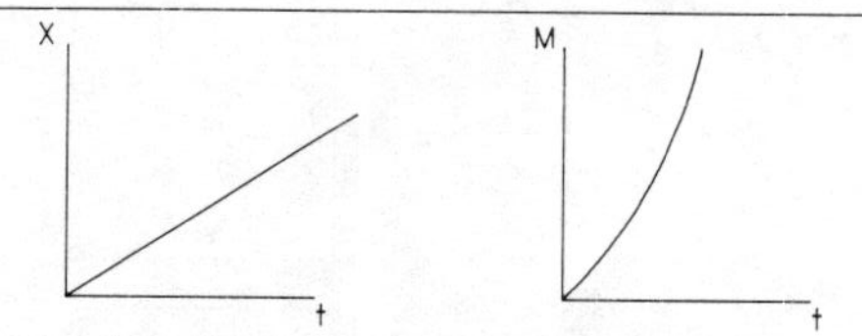	THE OUTPUT EQUALS THE INPUT RAISED TO A POWER (I.E., SECOND, THIRD, FOURTH, ETC.).
11	NONLINEAR OR UNSPECIFIED FUNCTION	f(X)	$M = f(X)$	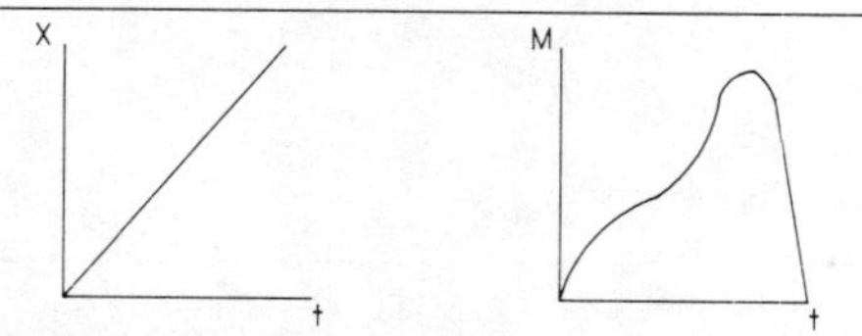	THE OUTPUT EQUALS SOME NONLINEAR OR UNSPECIFIED FUNCTION OF THE INPUT.
12	TIME FUNCTION	f(t)	$M = Xf(t)$ $M = f(t)$	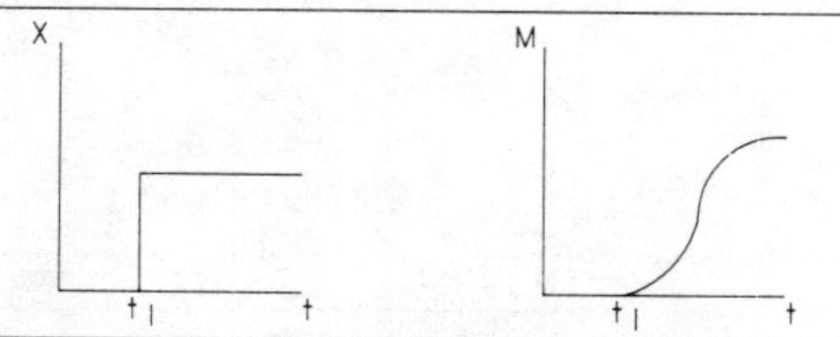	THE OUTPUT EQUALS THE INPUT TIMES SOME FUNCTION OF TIME OR EQUALS SOME FUNCTION OF TIME ALONE.
13	HIGH SELECTING	>	$M = \begin{cases} X_1 \text{ FOR } X_1 \geq X_2 \\ X_2 \text{ FOR } X_1 \leq X_2 \end{cases}$	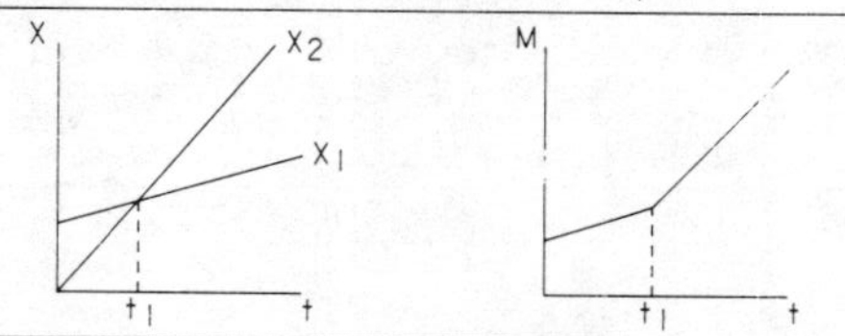	THE OUTPUT IS EQUAL TO THE GREATER OF THE INPUTS.

5.4 TABLE 3 - CONTINUED

NO	FUNCTION	SYMBOL	MATH EQUATION	GRAPHIC REPRESENTATION	DEFINITION
14	LOW SELECTING	<	$M = \begin{cases} X_1 \text{ FOR } X_1 \le X_2 \\ X_2 \text{ FOR } X_1 \ge X_2 \end{cases}$	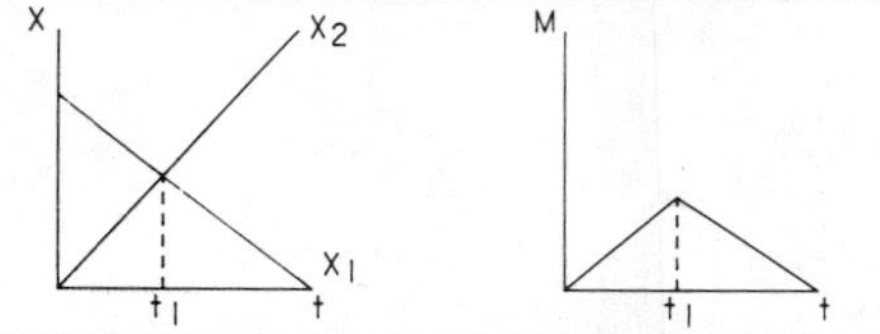	THE OUTPUT IS EQUAL TO THE LESSER OF THE INPUTS.
15	HIGH LIMITING	≯	$M = \begin{cases} X \text{ FOR } X \le H \\ H \text{ FOR } X \ge H \end{cases}$	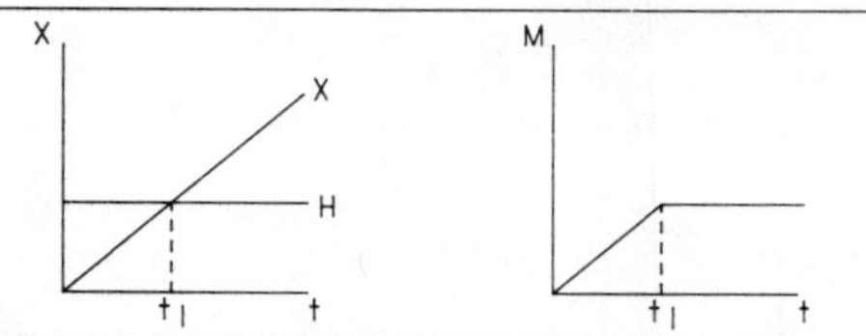	THE OUTPUT EQUALS THE INPUT OR THE HIGH LIMIT VALUE WHICHEVER IS LOWER.
16	LOW LIMITING	≮	$M = \begin{cases} X \text{ FOR } X \ge L \\ L \text{ FOR } X \le L \end{cases}$	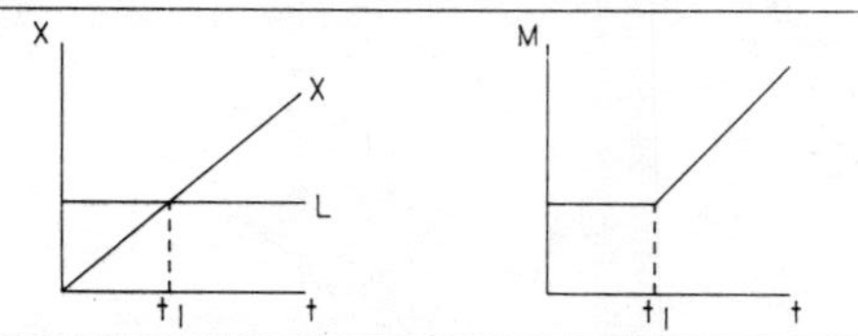	THE OUTPUT EQUALS THE INPUT OR THE LOW LIMIT VALUE WHICHEVER IS HIGHER.
17	REVERSE PROPORTIONAL	-K	$M = -KX$	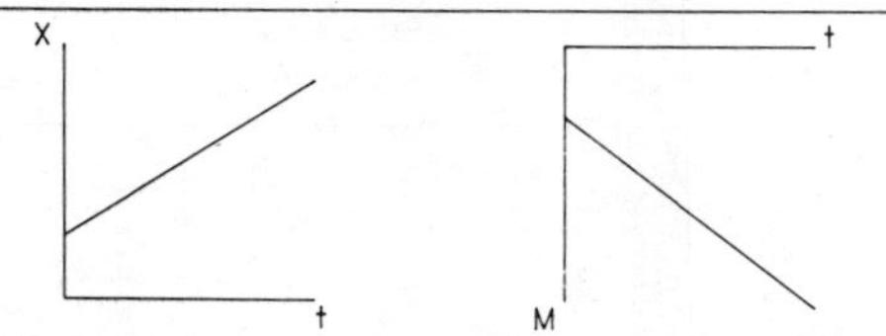	THE OUTPUT IS REVERSELY PROPORTIONAL TO THE INPUT.
18	VELOCITY LIMITER	∀̸	$\frac{dM}{dt} = \frac{dX}{dt} \begin{cases} \frac{dX}{dt} \le H \text{ AND} \\ M = X \end{cases}$ $\frac{dM}{dt} = H \begin{cases} \frac{dX}{dt} \ge H \text{ OR} \\ M \ne X \end{cases}$	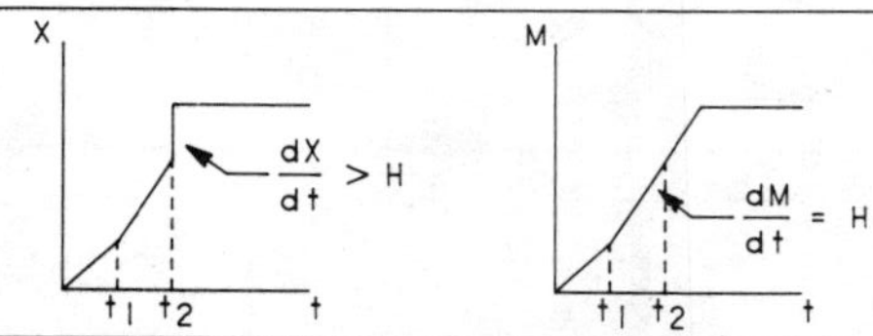	THE OUTPUT EQUALS THE INPUT AS LONG AS THE RATE OF CHANGE OF THE INPUT DOES NOT EXCEED A LIMIT VALUE. THE OUTPUT WILL CHANGE AT THE RATE ESTABLISHED BY THIS LIMIT UNTIL THE OUTPUT AGAIN EQUALS THE INPUT.
19	BIAS	+ − ±	$M = X \pm b$	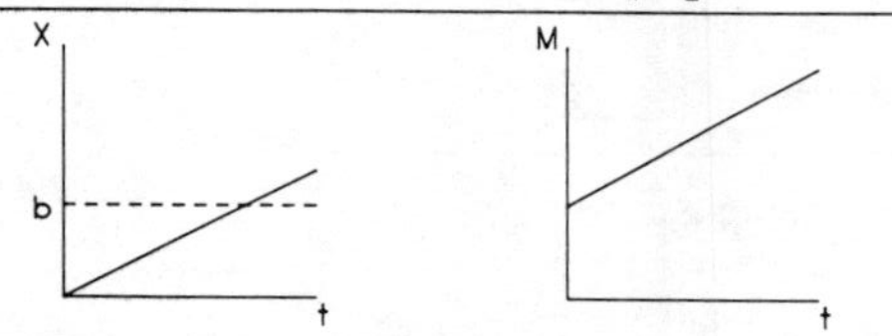	THE OUTPUT EQUALS THE INPUT PLUS (OR MINUS) SOME ARBITRARY VALUE (BIAS).
20	CONVERT	*/*	OUTPUT = f(INPUT)	NONE	THE FORM OF THE OUTPUT SIGNAL IS DIFFERENT FROM THAT OF THE INPUT. * E - VOLTAGE I - CURRENT P - PNEUMATIC A - ANALOG B - BINARY * H - HYDRAULIC O - ELECTROMAGNETIC, SONIC R - RESISTANCE(ELECT.) D - DIGITAL

5.4 TABLE 3 - CONTINUED

NO	FUNCTION	SYMBOL	MATH EQUATION	GRAPHIC REPRESENTATION	DEFINITION
21	SIGNAL MONITOR	**H	STATE 1 $X \leq H$ STATE 2 (ENERGIZED OR ALARM STATE) $X > H$	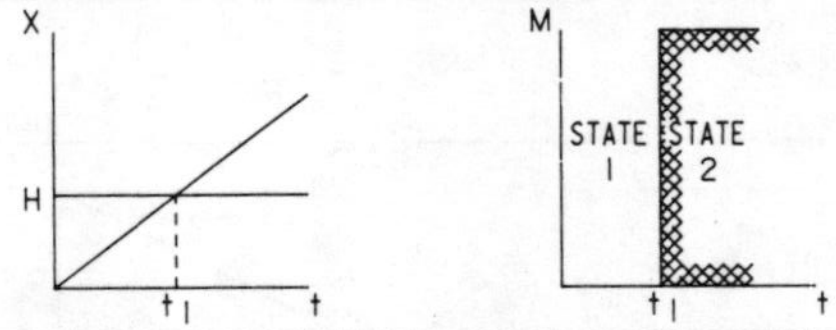 	THE OUTPUT HAS DISCRETE STATES WHICH ARE DEPENDENT ON THE VALUE OF THE INPUT. WHEN THE INPUT EXCEEDS (OR BECOMES LESS THAN) AN ARBITRARY LIMIT VALUE THE OUTPUT CHANGES STATE.
		**L	STATE 1 (ENERGIZED OR ALARM STATE) $X < L$ STATE 2 $X \geq L$	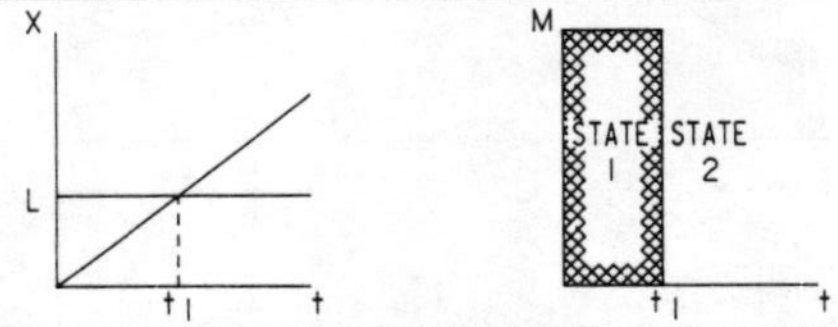 	
		**HL	STATE 1 (FIRST OUTPUT M_1 ENERGIZED OR ALARM STATE) $X < L$ STATE 2 (BOTH OUTPUTS INACTIVE OR DE-ENERGIZED) $L \leq X \leq H$ STATE 3 (SECOND OUTPUT M_2 ENERGIZED OR ALARM STATE) $X > H$	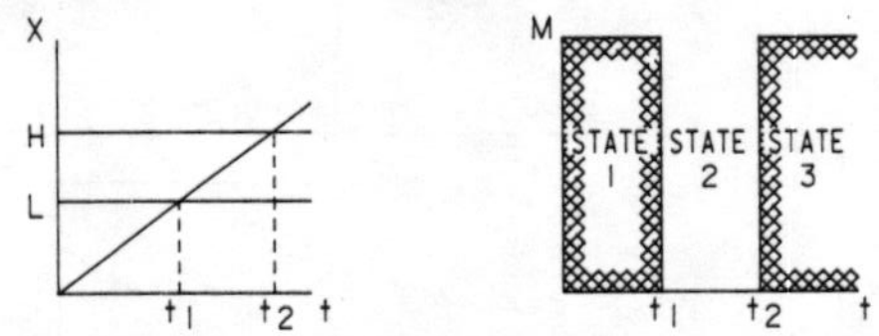 	

THE VARIABLES USED IN THE TABLE ARE:

- b - ANALOG BIAS VALUE.
- $\frac{d}{dt}$ - DERIVATIVE WITH RESPECT TO TIME.
- H - AN ARBITRARY ANALOG HIGH LIMIT VALUE.
- $\frac{1}{T_I}$ - INTEGRATING RATE.
- L - AN ARBITRARY ANALOG LOW LIMIT VALUE.
- M - ANALOG OUTPUT VARIABLE.
- n - NUMBER OF ANALOG INPUTS OR VALUE OF EXPONENT.
- t - TIME.
- T_D - DERIVATIVE TIME.
- X - ANALOG INPUT VARIABLE.
- $X_1, X_2, X_3, \ldots, X_n$ - ANALOG INPUT VARIABLE (1 TO N IN NUMBER).
- * - TABLE 1 LETTER DESIGNATORS.

NOTE: THE SQUARE MAY BE USED AS A FLAG

1-0	ON-OFF
REV	REVERSE ACTION

THIS TABLE HAS BEEN MODIFIED FROM SAMA PMC 22-11-1981 WITH PERMISSION OF THE COPYRIGHT HOLDER, SCIENTIFIC APPARATUS MAKERS ASSOCIATION. COPIES OF PMC 22-11-1981 ARE AVAILABLE FROM SAMA, 1101 16TH STREET NW, WASHINGTON, D.C., 20036.

6 DRAWINGS

6.1 Cautionary Notes

(1) If a given drawing, or set of drawings, uses graphic symbols that are similar or identical in shape or configuration and that have different meanings because they are taken from different standards, then adequate steps must be taken to avoid misinterpretation of the symbols used. These steps may be to use caution notes, reference notes, comparison charts that illustrate and define the conflicting symbols, or other suitable means. This requirement is especially critical in cases where symbols taken from different disciplines are intermixed and their misinterpretation might cause danger to personnel or damage to equipment.

(2) The titles *Simplified Diagrams, Conceptual Diagrams* and *Detailed Diagrams* of Section 6.12 were chosen to represent a cross section of symbol usage, not any particular generic document. (*See* 4.4.10 for a more complete discussion.)

(3) The line symbols of Section 6.2 offer ''user's choice'' alternative electrical symbols and optional binary symbols. The subsequent examples use one consistent set of these alternatives and apply the binary options. This was done for consistency of appearance of the standard.

It is recommended that the user choose either the dashed line electrical symbol or the triple cross hatch symbol and apply it consistently. The optional binary (on-off) symbols are available for those applications where the user finds it necessary to distinguish between analog and binary signals. If, in the user's judgment, the application does not require such differentiation, the reverse slash may be omitted from on-off signal line symbols. Consistency is recommended on a given set of documents.

6.2 INSTRUMENT LINE SYMBOLS

ALL LINES TO BE FINE IN RELATION TO PROCESS PIPING LINES.

(1)	INSTRUMENT SUPPLY * OR CONNECTION TO PROCESS	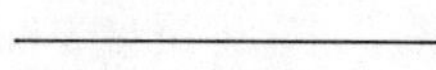
(2)	UNDEFINED SIGNAL	
(3)	PNEUMATIC SIGNAL **	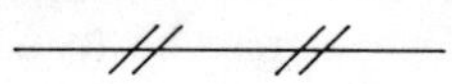
(4)	ELECTRIC SIGNAL	
(5)	HYDRAULIC SIGNAL	
(6)	CAPILLARY TUBE	
(7)	ELECTROMAGNETIC OR SONIC SIGNAL *** (GUIDED)	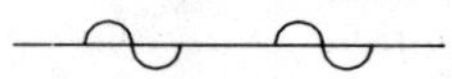
(8)	ELECTROMAGNETIC OR SONIC SIGNAL *** (NOT GUIDED)	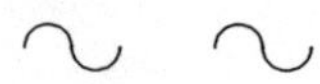
(9)	INTERNAL SYSTEM LINK (SOFTWARE OR DATA LINK)	
(10)	MECHANICAL LINK	

OPTIONAL BINARY (ON-OFF) SYMBOLS

(11)	PNEUMATIC BINARY SIGNAL	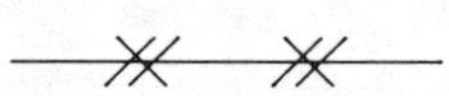
(12)	ELECTRIC BINARY SIGNAL	

NOTE: 'Or' means user's choice. Consistency is recommended.

* The following abbreviations are suggested to denote the types of power supply. These designations may also be applied to purge fluid supplies.

AS - Air Supply
- IA - Instrument Air } Options
- PA - Plant Air } Options

ES - Electric Supply
GS - Gas Supply
HS - Hydraulic Supply
NS - Nitrogen Supply
SS - Steam Supply
WS - Water Supply

The supply level may be added to the instrument supply line, e.g., AS-100, a 100-psig air supply; ES-24DC, a 24-volt direct current power supply.

** The pneumatic signal symbol applies to a signal using any gas as the signal medium. If a gas other than air is used, the gas may be identified by a note on the signal symbol or otherwise.

*** Electromagnetic phenomena include heat, radio waves, nuclear radiation, and light.

6.3 GENERAL INSTRUMENT OR FUNCTION SYMBOLS

	PRIMARY LOCATION *** NORMALLY ACCESSIBLE TO OPERATOR	FIELD MOUNTED	AUXILIARY LOCATION *** NORMALLY ACCESSIBLE TO OPERATOR
DISCRETE INSTRUMENTS	1 * IPI ** 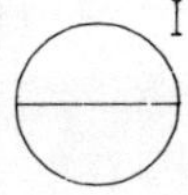	2 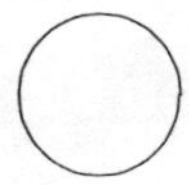	3 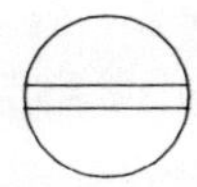
SHARED DISPLAY, SHARED CONTROL	4 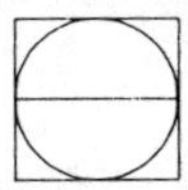	5 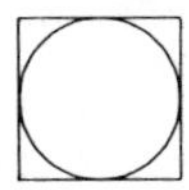	6 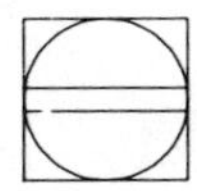
COMPUTER FUNCTION	7 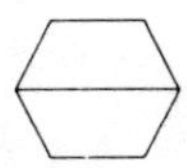	8 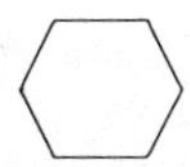	9 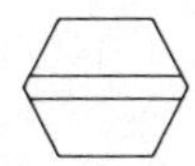
PROGRAMMABLE LOGIC CONTROL	10 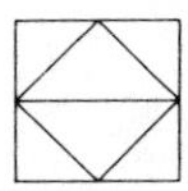	11 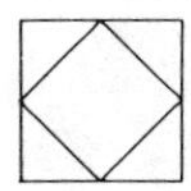	12 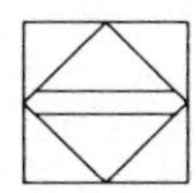

* Symbol size may vary according to the user's needs and the type of document. A suggested square and circle size for large diagrams is shown above. Consistency is recommended.

** Abbreviations of the user's choice such as IPI (Instrument Panel #1), IC2 (Instrument Console #2), CC3 (Computer Console #3), etc., may be used when it is necessary to specify instrument or function location.

*** Normally inaccessible or behind-the-panel devices or functions may be depicted by using the same symbols but with dashed horizontal bars, i.e.

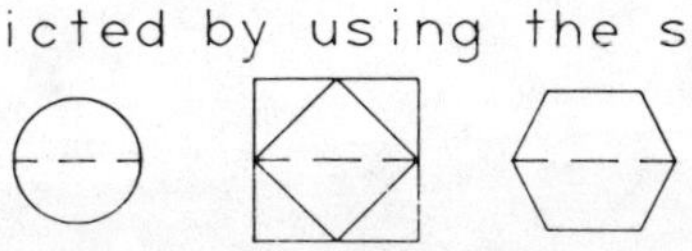

6.3 GENERAL INSTRUMENT OR FUNCTION SYMBOLS (Contd.)

13	14	15
	 INSTRUMENT WITH LONG TAG NUMBER	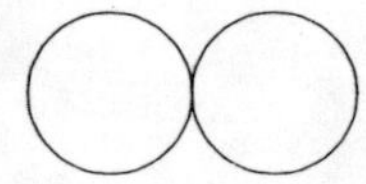 INSTRUMENTS SHARING COMMON HOUSING *
16  PILOT LIGHT	17 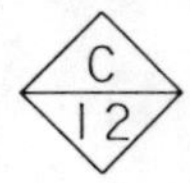 PANEL MOUNTED PATCHBOARD POINT 12	18 **  PURGE OR FLUSHING DEVICE
19 **  RESET FOR LATCH-TYPE ACTUATOR	20 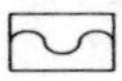 DIAPHRAGM SEAL	21 ** ***  UNDEFINED INTERLOCK LOGIC

* It is not mandatory to show a common housing.

** These diamonds are approximately half the size of the larger ones.

*** For specific logic symbols, see ANSI/ISA Standard S5.2.

6.4 CONTROL VALVE BODY SYMBOLS, DAMPER SYMBOLS

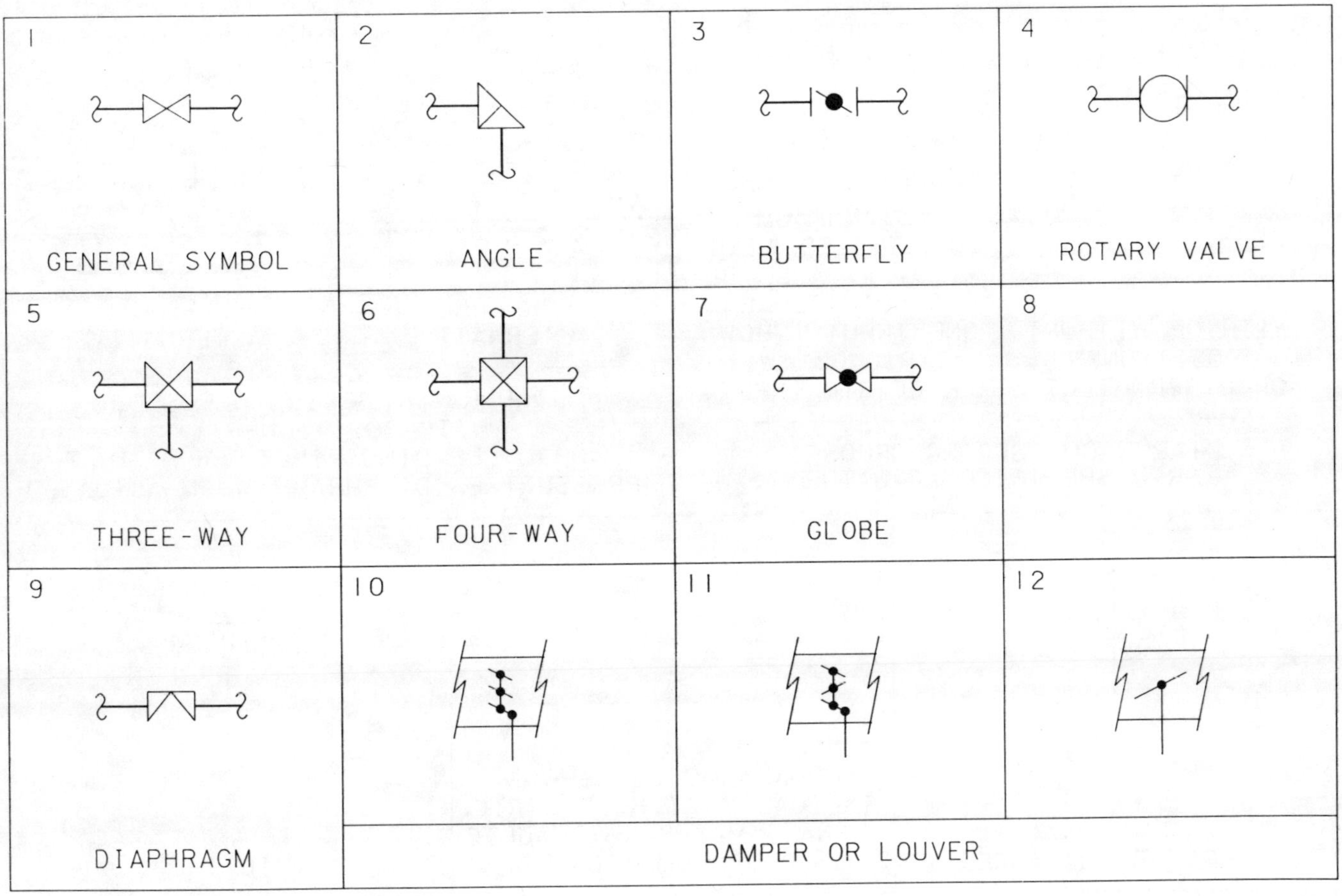

Further information may be added adjacent to the body symbol either by note or code number.

6.5 ACTUATOR SYMBOLS

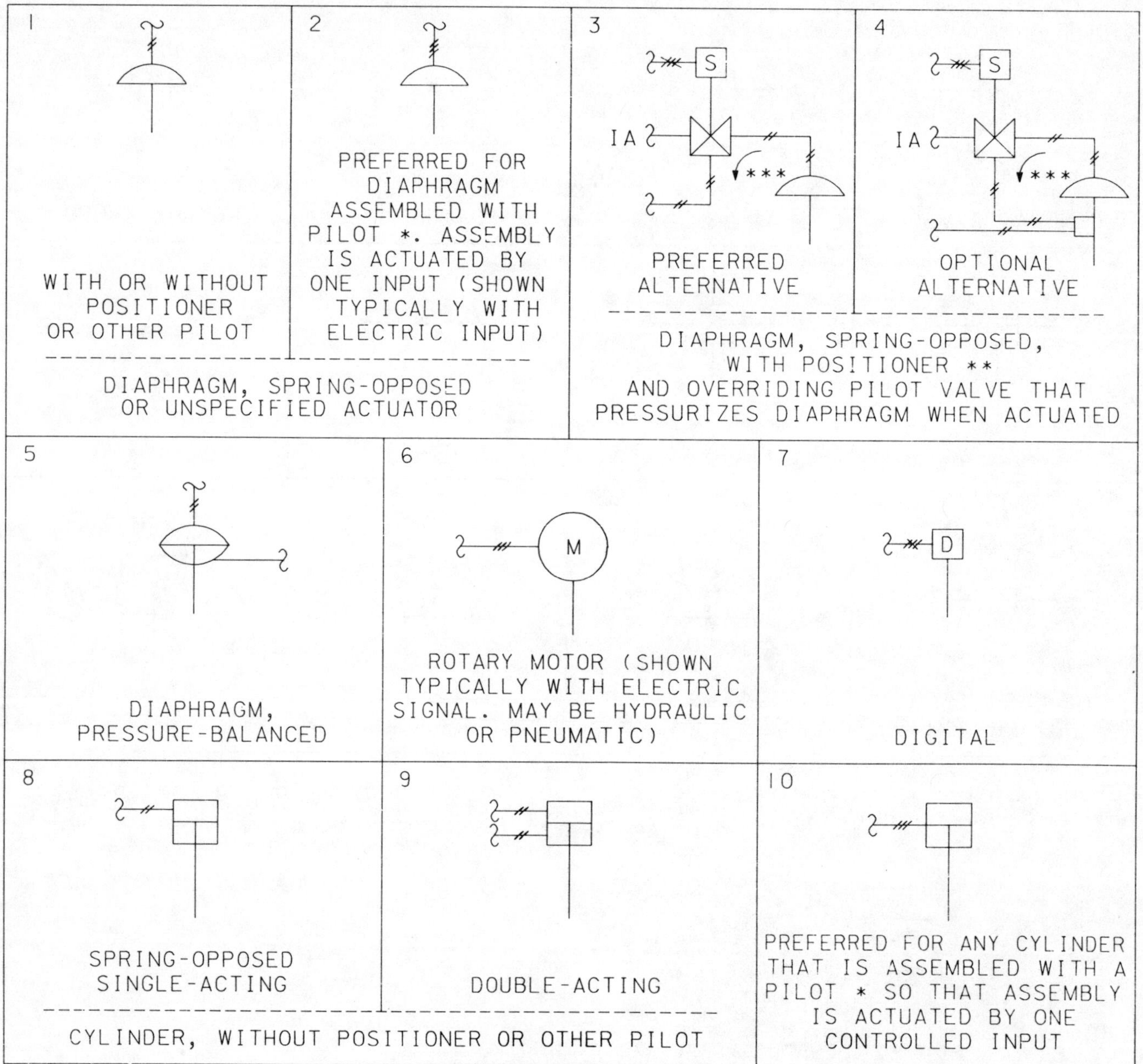

* Pilot may be positioner, solenoid valve, signal converter, etc.

** The positioner need not be shown unless an intermediate device is on its output. The positioner tagging, ZC, need not be used even if the positioner is shown. The positioner symbol, a box drawn on the actuator shaft, is the same for all types of actuators. When the symbol is used, the type of instrument signal, i.e., pneumatic, electric, etc., is drawn as appropriate. If the positioner symbol is used and there is no intermediate device on its output, then the positioner output signal need not be shown.

*** The arrow represents the path from a common to a fail open port. It does not correspond necessarily to the direction of fluid flow.

6.5 ACTUATOR SYMBOLS (Contd.)

11

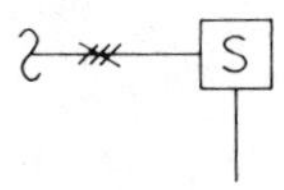

SOLENOID

12

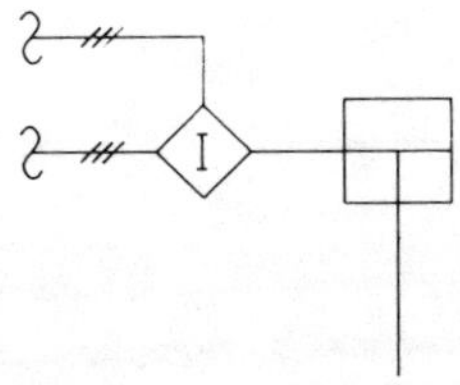

PREFERRED ALTERNATIVE. A BUBBLE WITH INSTRUMENT TAGGING, E.G., TY-1, MAY BE USED INSTEAD OF THE INTERLOCK SYMBOL ◇I

13

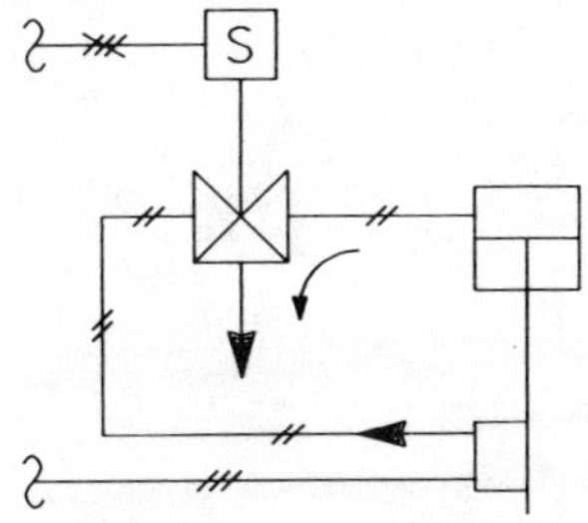

SINGLE-ACTING CYLINDER (IMPLIED I/P)

CYLINDER WITH POSITIONER AND OVERRIDING PILOT VALVE

14

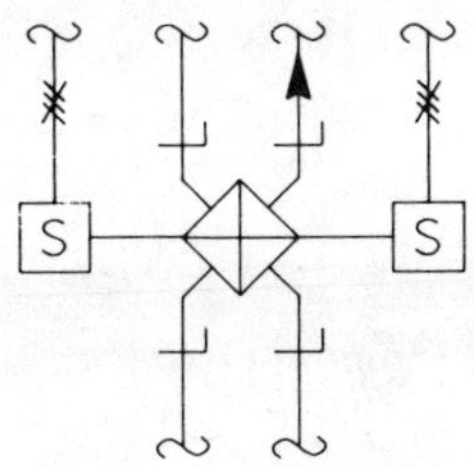

DUAL SOLENOIDS SWITCHING 4-WAY HYDRAULIC VALVE

15

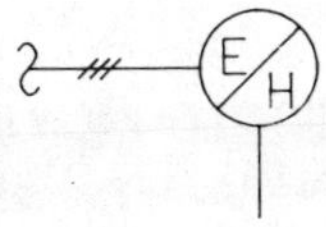

ELECTROHYDRAULIC

16

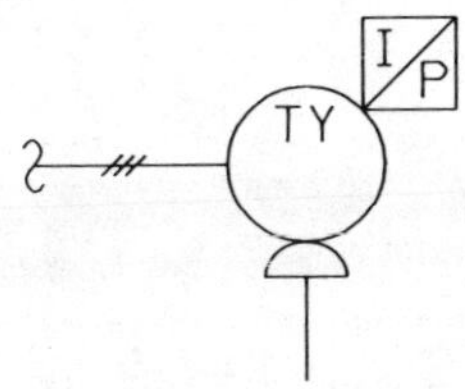

VALVE ACTUATOR WITH ATTACHED ELECTRO-PNEUMATIC CONVERTER

17

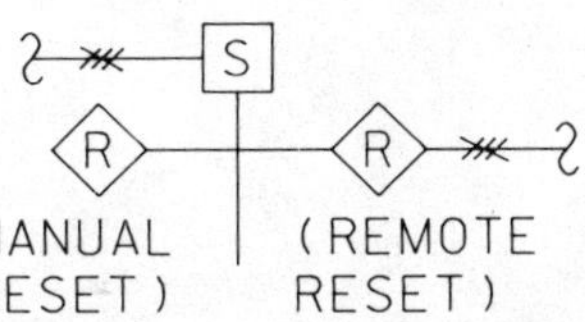

LATCH-TYPE ACTUATOR WITH RESET (SHOWN TYPICALLY FOR SOLENOID ACTUATOR AND TYPICALLY WITH ELECTRIC SIGNAL FOR REMOTE RESET, WITH MANUAL RESET ALTERNATIVE)

18

FOR PRESSURE RELIEF OR SAFETY VALVES ONLY: DENOTES A SPRING, WEIGHT, OR INTEGRAL PILOT

19

HAND ACTUATOR OR HANDWHEEL

6.6 SYMBOLS FOR SELF-ACTUATED REGULATORS, VALVES, AND OTHER DEVICES

FLOW	1 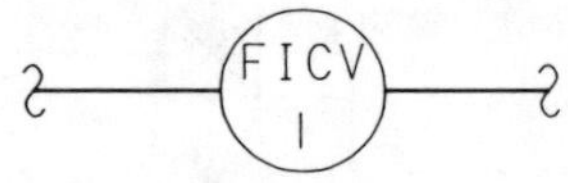 AUTOMATIC REGULATOR WITH INTEGRAL FLOW INDICATION	2 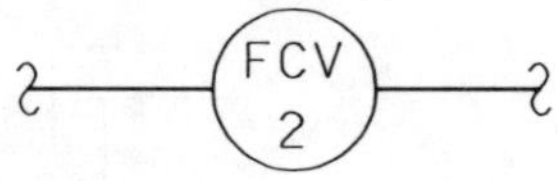 AUTOMATIC REGULATOR WITHOUT INDICATION	3 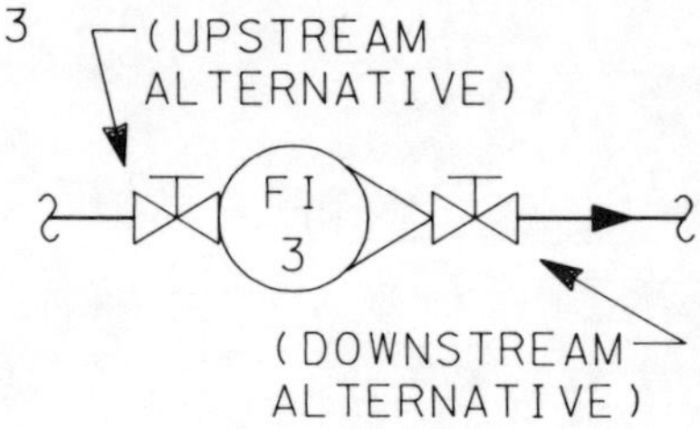 INDICATING VARIABLE AREA METER WITH INTEGRAL MANUAL THROTTLE VALVE
FLOW	4 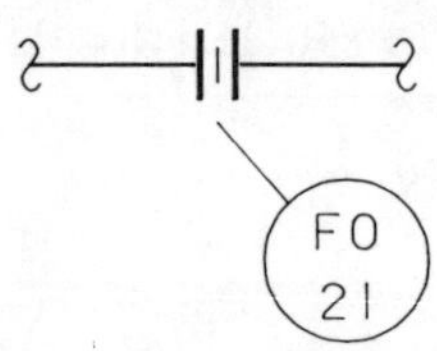 RESTRICTION ORIFICE (ORIFICE PLATE, CAPILLARY TUBE OR MULTI-STAGE TYPE, ETC.) IN PROCESS LINE	5 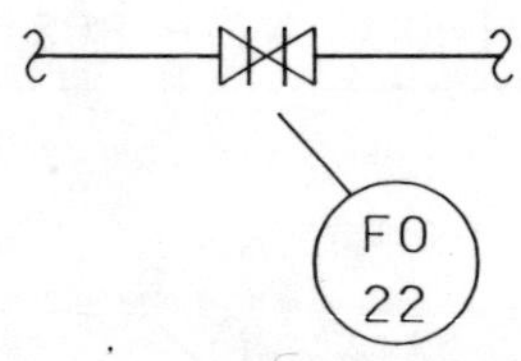 RESTRICTION ORIFICE DRILLED IN VALVE (INSTRUMENT TAG NUMBER MAY BE OMITTED IF VALVE IS OTHERWISE IDENTIFIED)	6 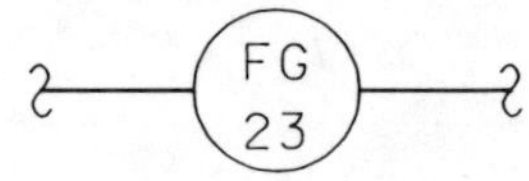 FLOW SIGHT GLASS, PLAIN OR WITH PADDLE WHEEL, FLAPPER, ETC.
FLOW	7 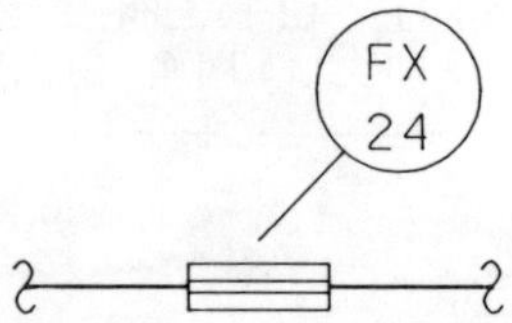 FLOW STRAIGHTENING VANE (USE OF TAG NUMBER IS OPTIONAL. THE LOOP NUMBER MAY BE THE SAME AS THAT OF THE ASSOCIATED PRIMARY ELEMENT)	8	9
HAND	1 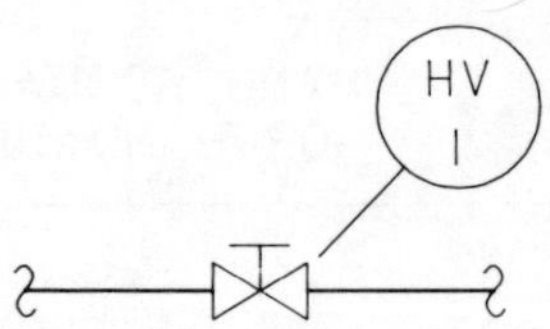 HAND CONTROL VALVE IN PROCESS LINE	2 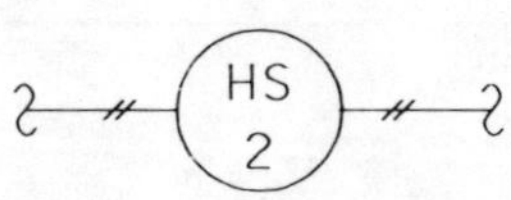 HAND-ACTUATED ON-OFF SWITCHING VALVE IN PNEUMATIC SIGNAL LINE	3 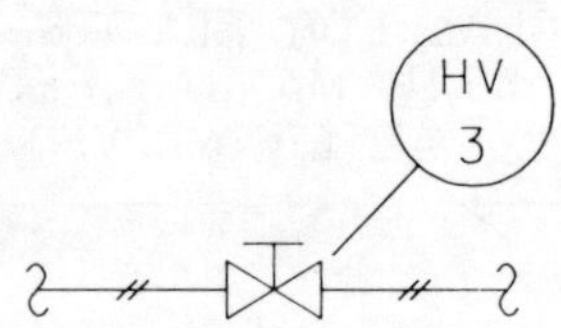 HAND CONTROL VALVE IN SIGNAL LINE

6.6 SYMBOLS FOR SELF-ACTUATED REGULATORS, VALVES, AND OTHER DEVICES (Contd.)

LEVEL

1

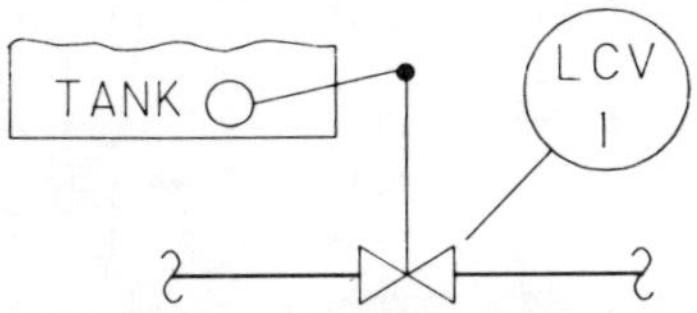

LEVEL REGULATOR WITH MECHANICAL LINKAGE

2

3

PRESSURE

1

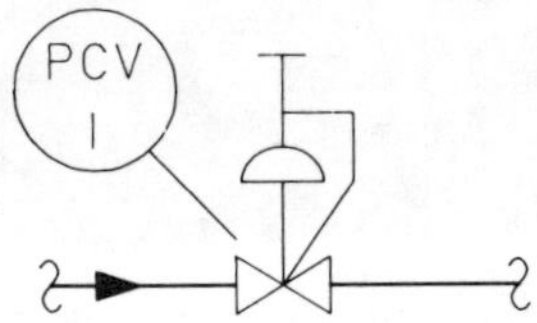

PRESSURE-REDUCING REGULATOR, SELF-CONTAINED, WITH HANDWHEEL ADJUSTABLE SET POINT

2

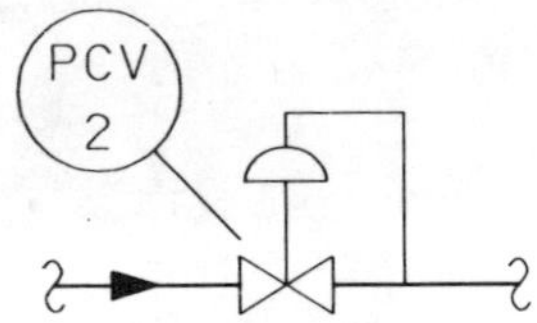

PRESSURE-REDUCING REGULATOR WITH EXTERNAL PRESSURE TAP

3

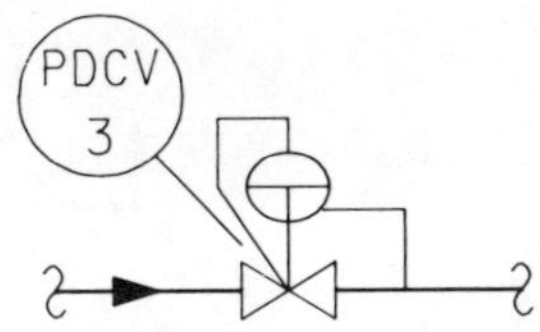

DIFFERENTIAL-PRESSURE-REDUCING REGULATOR WITH INTERNAL AND EXTERNAL PRESSURE TAPS

4

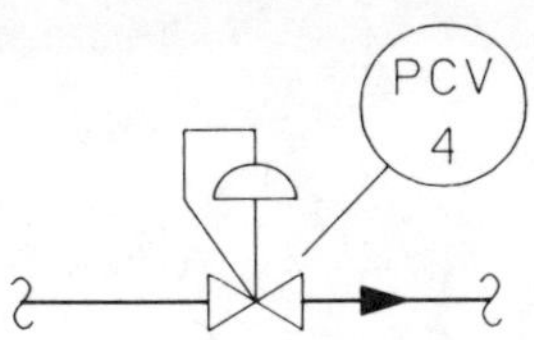

BACKPRESSURE REGULATOR, SELF-CONTAINED

5

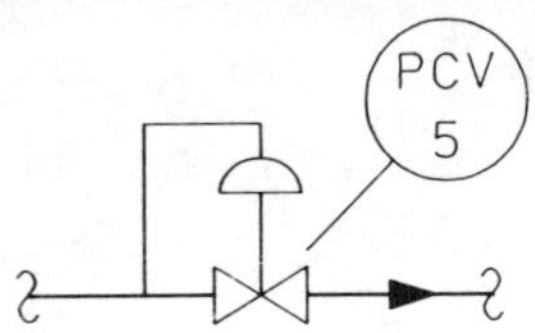

BACKPRESSURE REGULATOR WITH EXTERNAL PRESSURE TAP

6

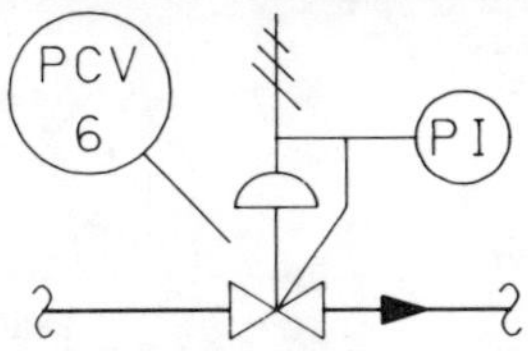

PRESSURE-REDUCING REGULATOR WITH INTEGRAL OUTLET PRESSURE RELIEF VALVE, AND OPTIONAL PRESSURE INDICATOR (TYPICAL AIR SET)

7

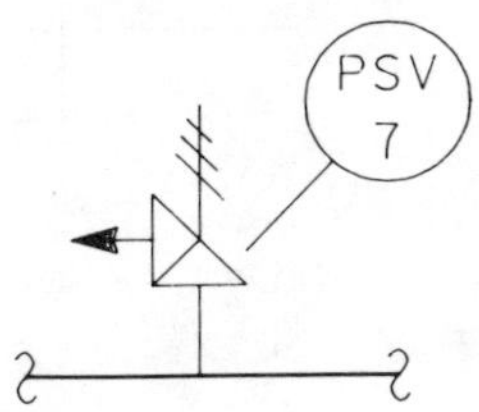

PRESSURE RELIEF OR SAFETY VALVE, GENERAL SYMBOL

8

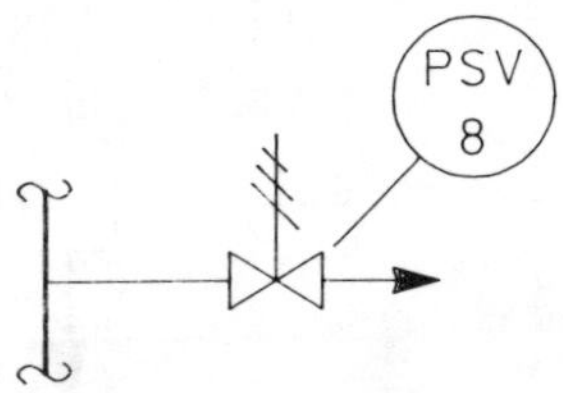

PRESSURE RELIEF OR SAFETY VALVE, STRAIGHT-THROUGH PATTERN, SPRING- OR WEIGHT-LOADED, OR WITH INTEGRAL PILOT

9

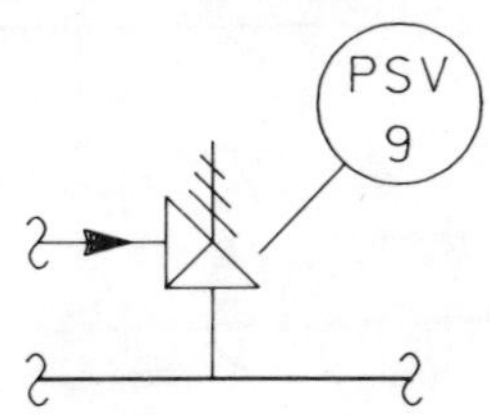

VACUUM RELIEF VALVE, GENERAL SYMBOL

6.6 SYMBOLS FOR SELF-ACTUATED REGULATORS, VALVES, AND OTHER DEVICES (Contd.)

PRESSURE (Contd.)	10 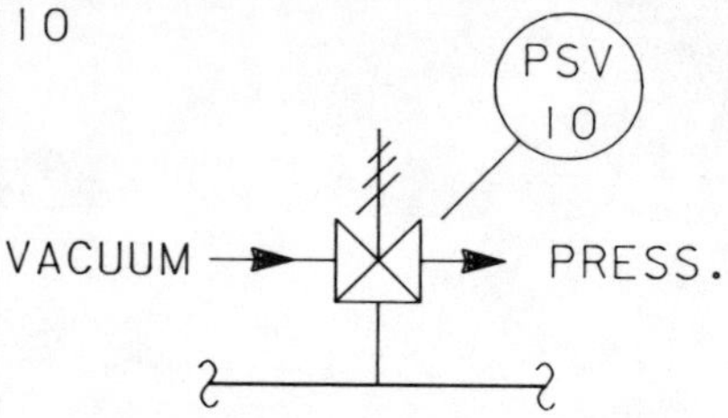 PRESSURE AND VACUUM RELIEF VALVE, SPRING- OR WEIGHT-LOADED, OR WITH INTEGRAL PILOT	11 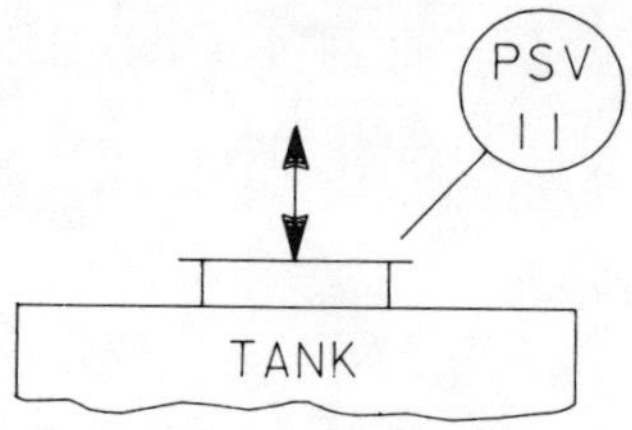 PRESSURE AND VACUUM RELIEF MANHOLE COVER	12 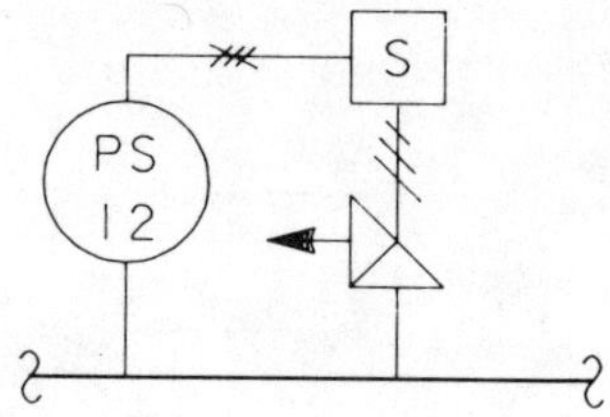 PRESSURE RELIEF OR SAFETY VALVE, ANGLE PATTERN, TRIPPED BY INTEGRAL SOLENOID *
	13 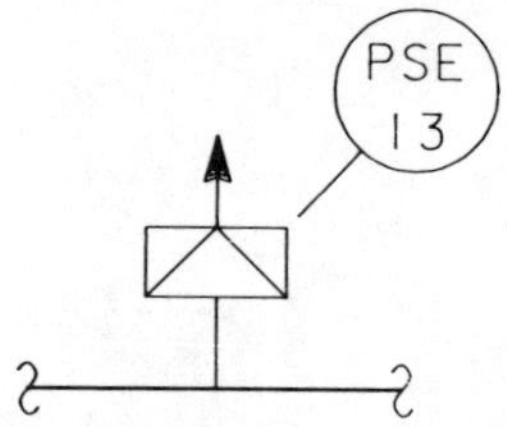 RUPTURE DISK OR SAFETY HEAD FOR PRESSURE RELIEF	14 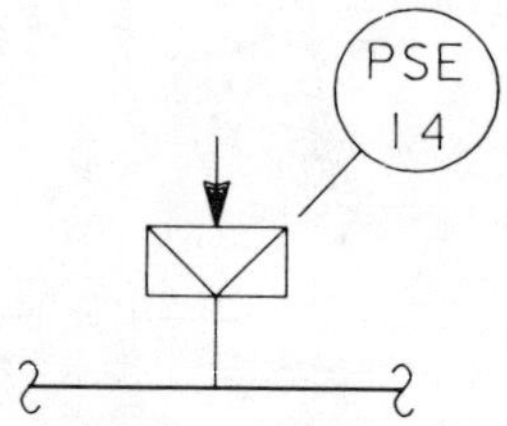 RUPTURE DISK OR SAFETY HEAD FOR VACUUM RELIEF	15 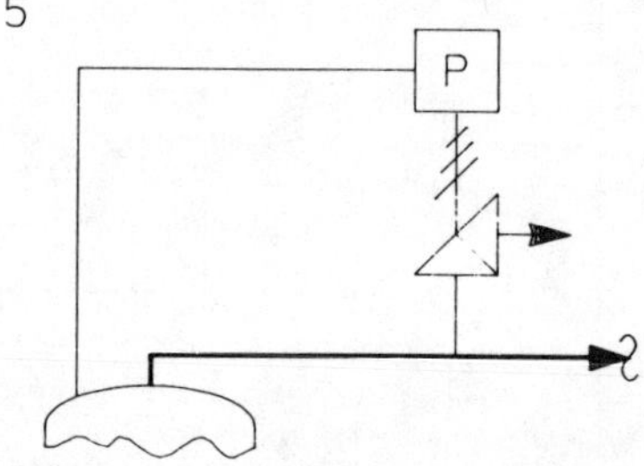 PILOT OPERATED RELIEF VALVE
TEMPERATURE	1 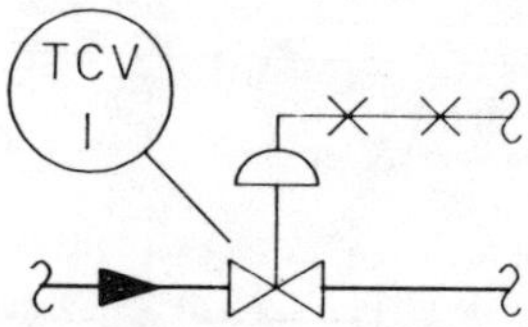 TEMPERATURE REGULATOR, FILLED-SYSTEM TYPE	2 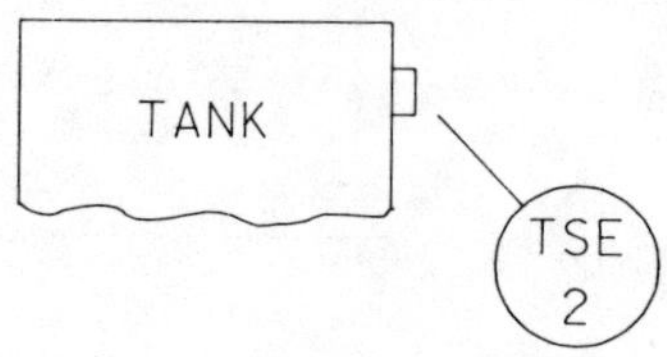 FUSIBLE PLUG OR DISK	3
TRAPS	1 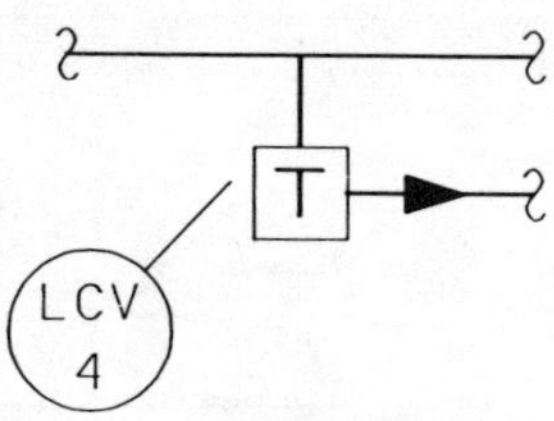 ALL TRAPS	2 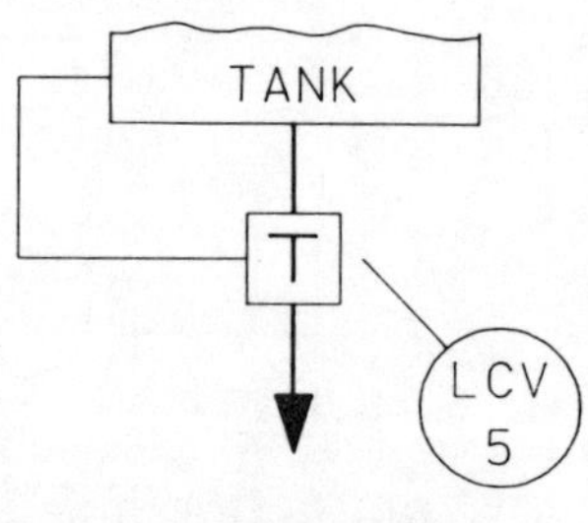 TRAP WITH EQUALIZING CONNECTION	3 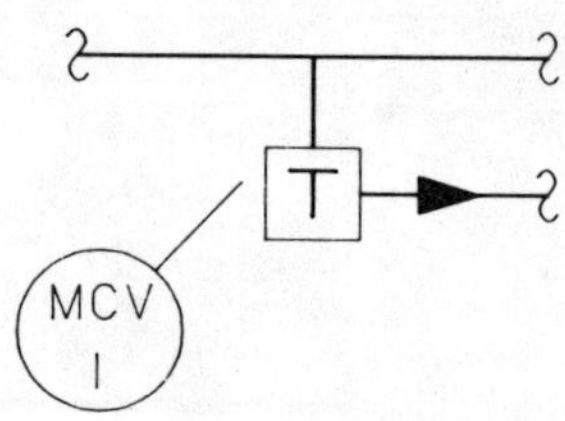 USER DEFINED TRAP

* The solenoid-tripped pressure relief valve is one of the class of power-actuated relief valves and is grouped with the other types of relief valves even though it is not entirely a self-actuated device.

6.7 SYMBOLS FOR ACTUATOR ACTION IN EVENT OF ACTUATOR POWER FAILURE. (SHOWN TYPICALLY FOR DIAPHRAGM-ACTUATED CONTROL VALVE)

1 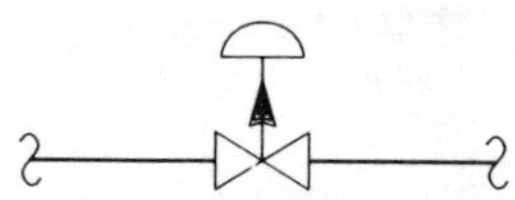 TWO-WAY VALVE, FAIL OPEN	2 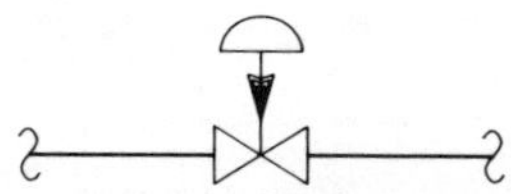 TWO-WAY VALVE, FAIL CLOSED	3 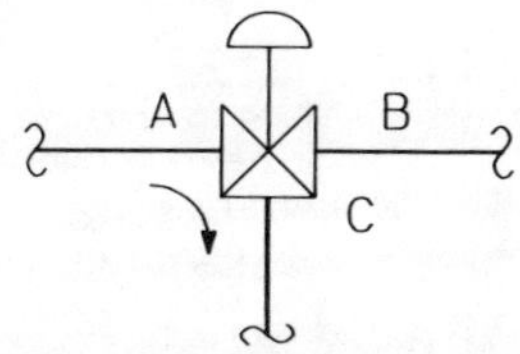 THREE-WAY VALVE, FAIL OPEN TO PATH A-C
4 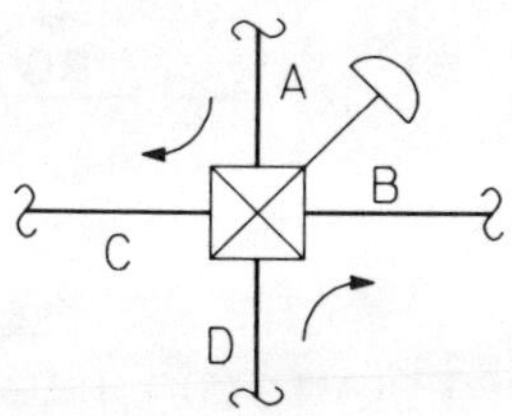 FOUR-WAY VALVE, FAIL OPEN TO PATHS A-C AND D-B	5 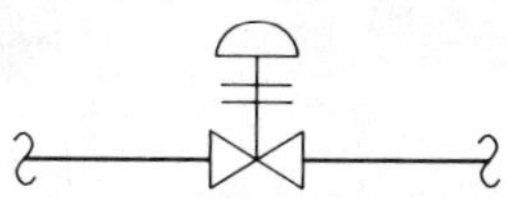 ANY VALVE, FAIL LOCKED (POSITION DOES NOT CHANGE)	6 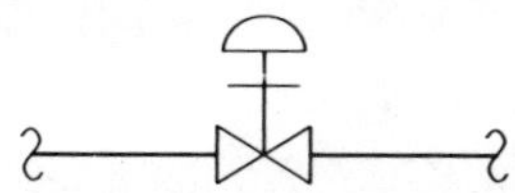 ANY VALVE, FAIL INDETERMINATE

The failure modes indicated are those commonly defined by the term, "shelf-position". As an alternative to the arrows and bars, the following abbreviations may be employed:

FO - Fail Open
FC - Fail Closed
FL - Fail Locked (last position)
FI - Fail Indeterminate

6.8 PRIMARY ELEMENT SYMBOLS

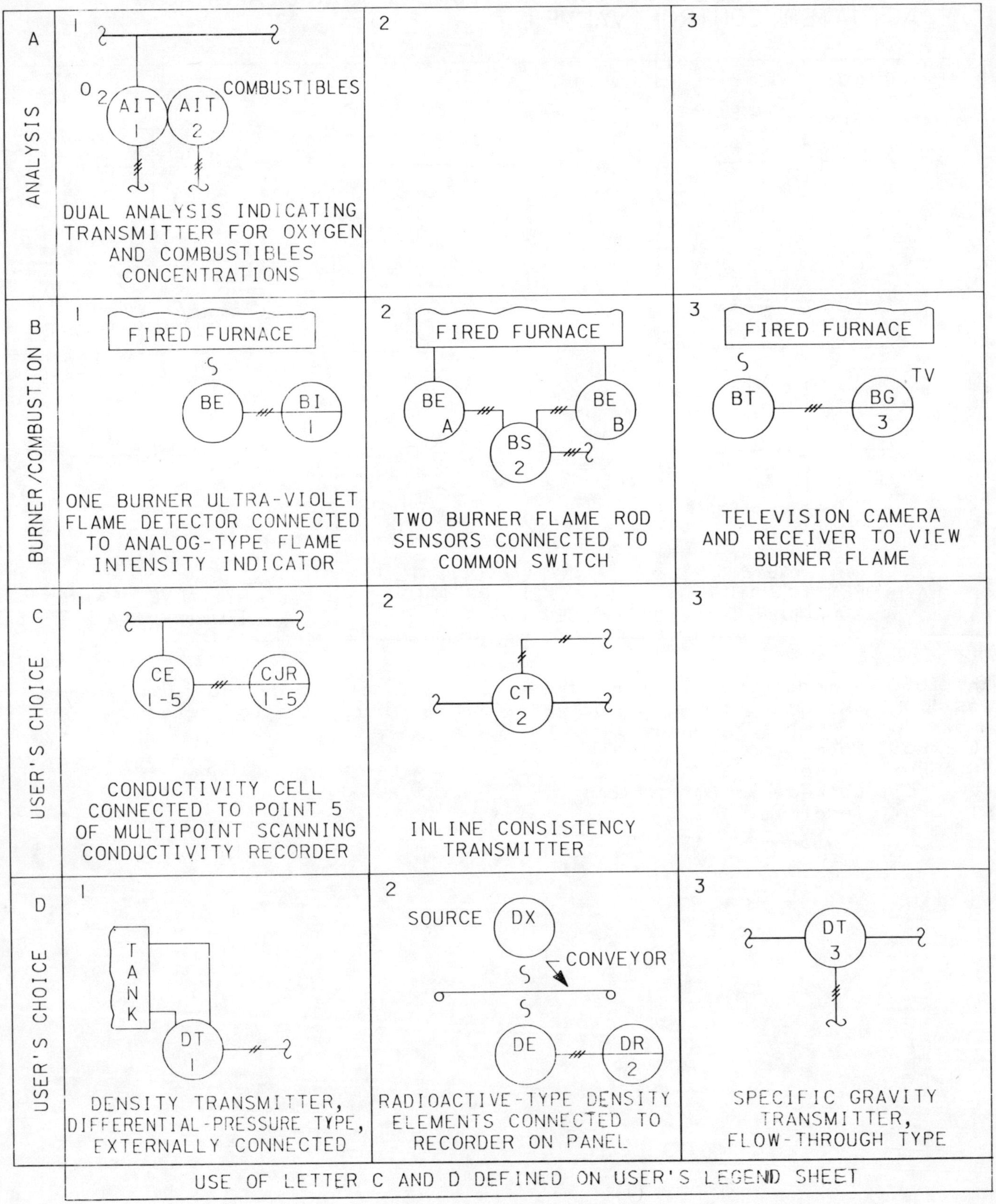

6.8 PRIMARY ELEMENT SYMBOLS (Contd.)

E VOLTAGE	1 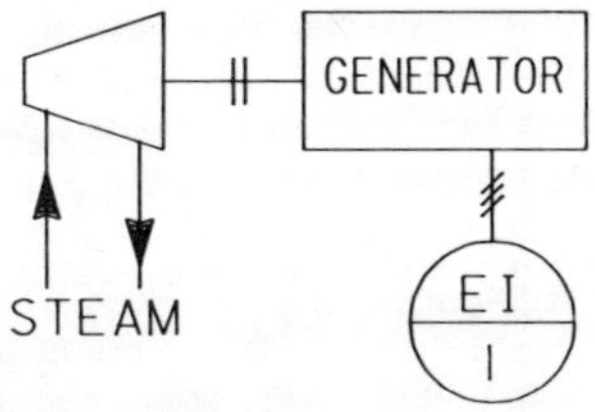 INDICATING VOLTMETER CONNECTED TO TURBINE-GENERATOR	2 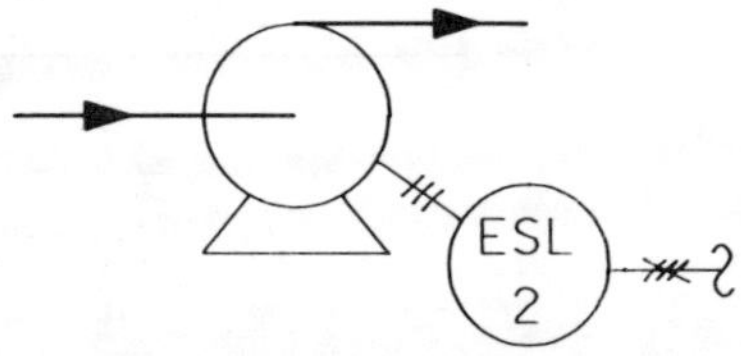 LOW-VOLTAGE SWITCH CONNECTED TO PUMP MOTOR	3
F FLOW RATE	1 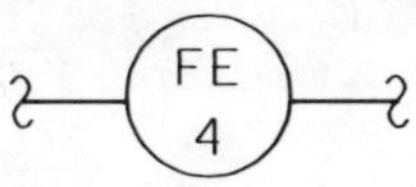 GENERAL SYMBOL THE WORDS LAMINAR, ETC., MAY BE ADDED	2 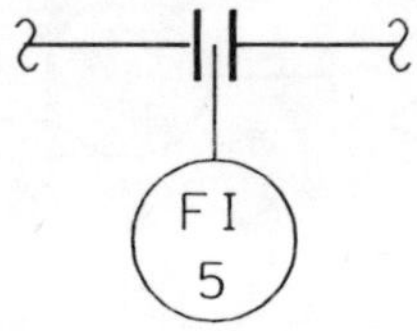 ORIFICE PLATE WITH FLANGE OR CORNER TAPS CONNECTED TO DIFFERENTIAL-PRESSURE TYPE FLOW INDICATOR	3 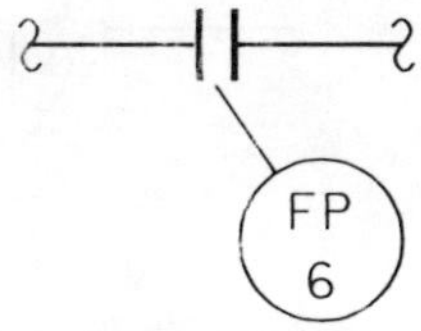 FLANGE OR CORNER TAP TEST CONNECTIONS WITHOUT ORIFICE PLATE
	4 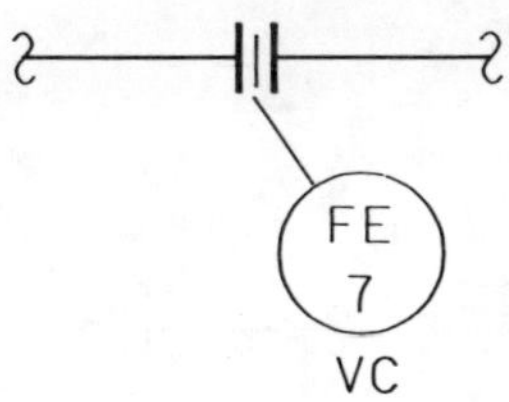 ORIFICE PLATE WITH VENA CONTRACTA TAPS	5 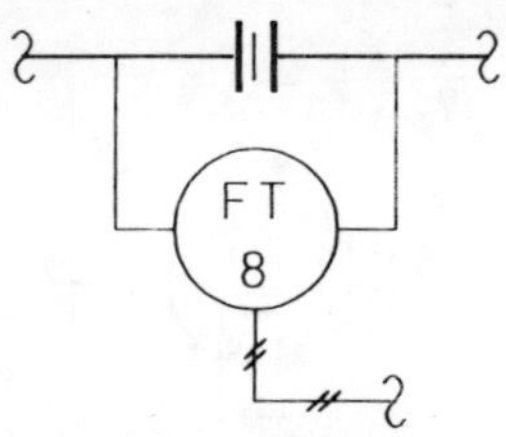 ORIFICE PLATE WITH VENA CONTRACTA, RADIUS, OR PIPE TAPS CONNECTED TO DIFFERENTIAL-PRESSURE-TYPE FLOW TRANSMITTER	6 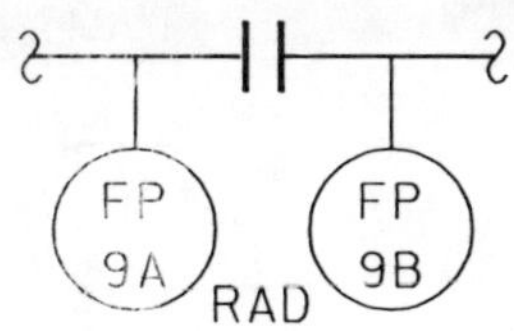 RADIUS TAP TEST CONNECTIONS WITHOUT ORIFICE PLATE
	7 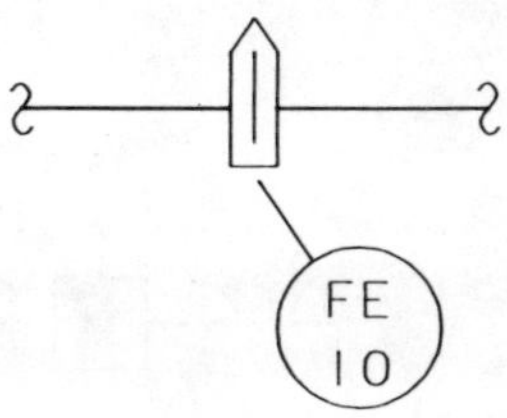 ORIFICE PLATE IN QUICK-CHANGE FITTING	8 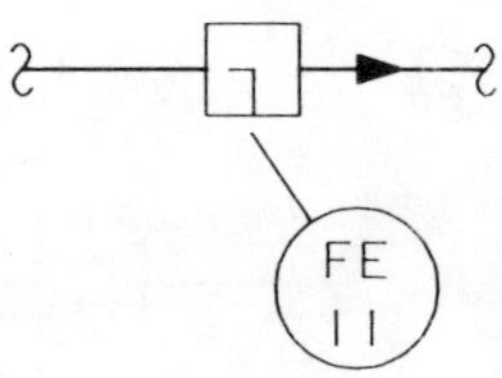 SINGLE PORT PITOT TUBE OR PITOT-VENTURI TUBE	9 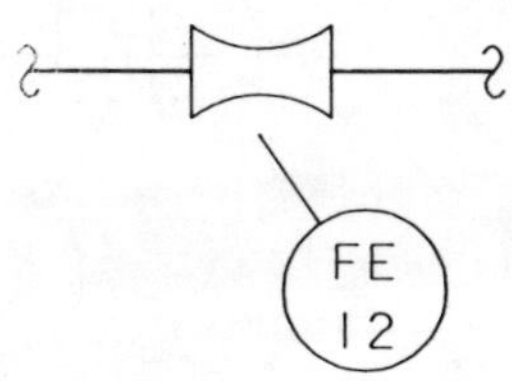 VENTURI TUBE

6.8 PRIMARY ELEMENT SYMBOLS (Contd.)

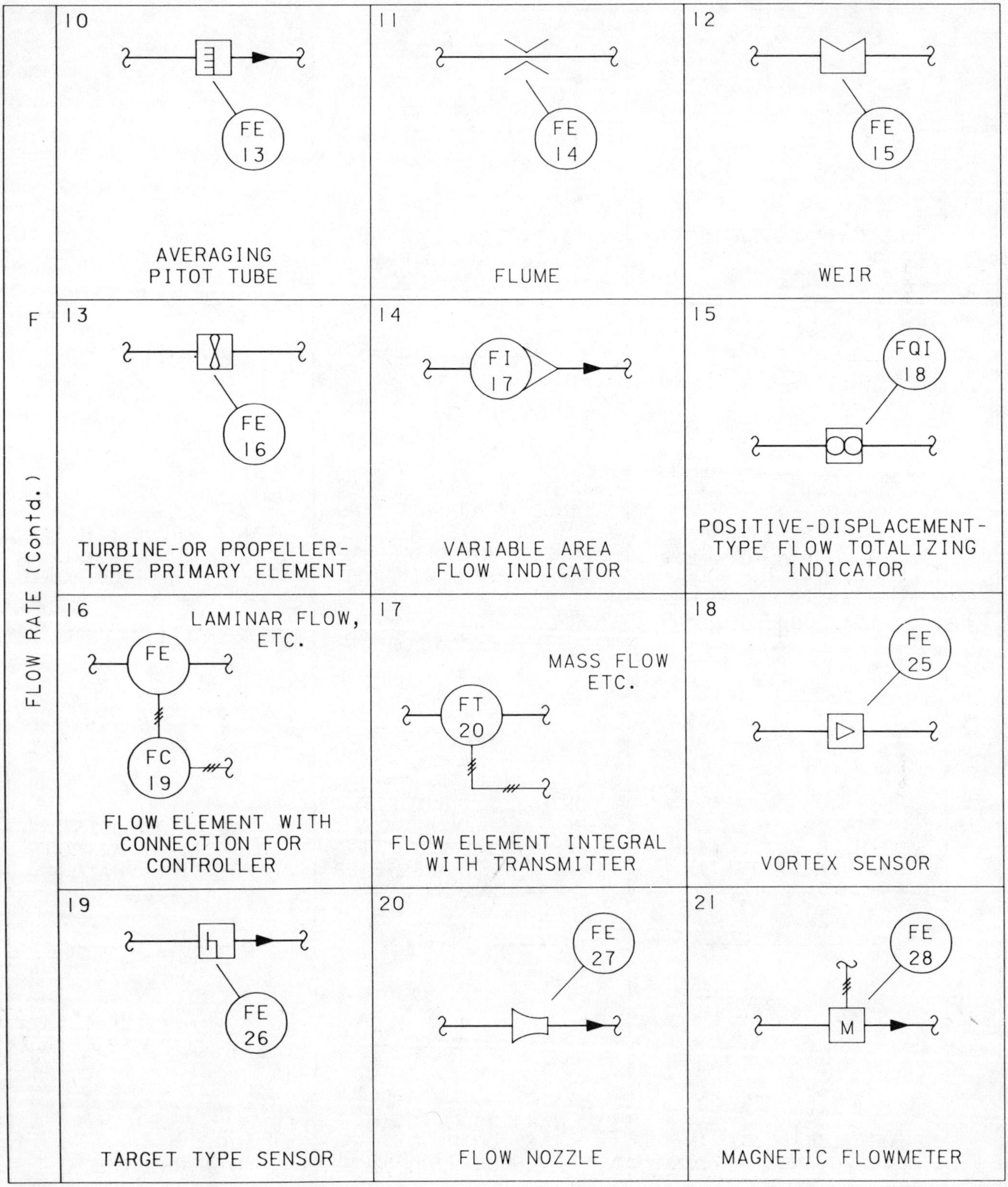

6.8 PRIMARY ELEMENT SYMBOLS (Contd.)

F FLOW RATE (Contd.)	22 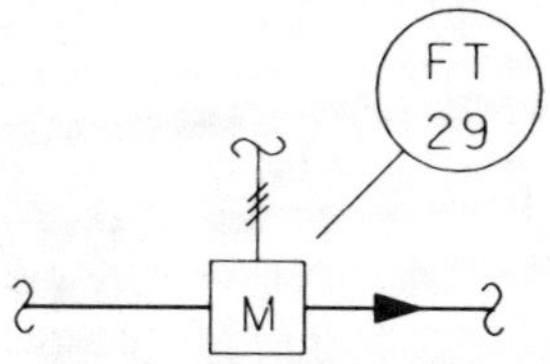 MAGNETIC FLOWMETER WITH INTEGRAL TRANSMITTER	23 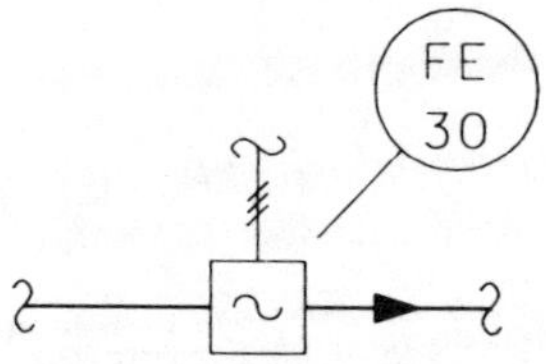 SONIC FLOWMETER "DOPPLER" OR "TRANSIT TIME" MAY BE ADDED	24
I CURRENT	1 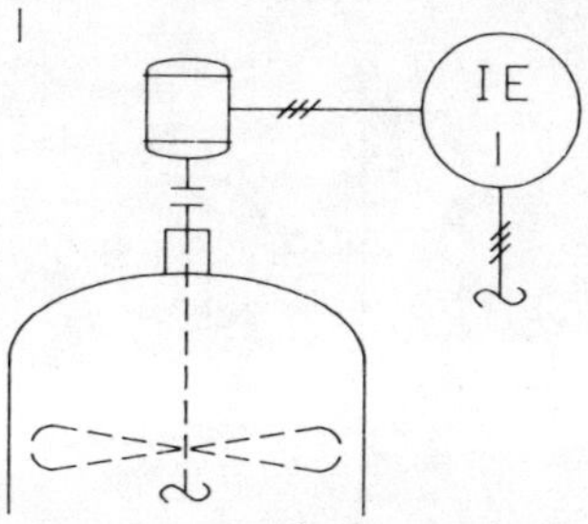 CURRENT TRANSFORMER MEASURING CURRENT OF ELECTRIC MOTOR	2	3
J POWER	1 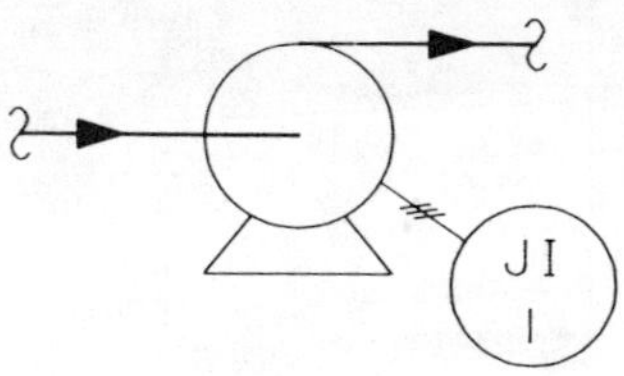 INDICATING WATTMETER CONNECTED TO PUMP MOTOR	2	3
K TIME OR TIME-SCHEDULE	1 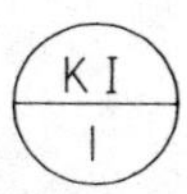 CLOCK	2 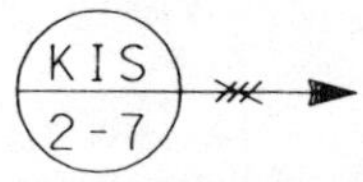 MULTIPOINT ON-OFF TIME SEQUENCING PROGRAMMER POINT 7	3 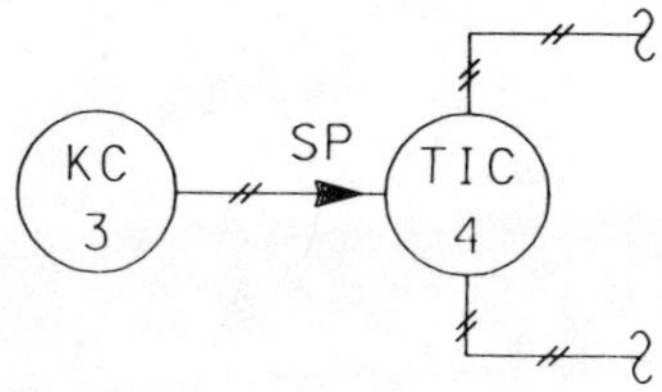 TIME-SCHEDULE CONTROLLER, ANALOG TYPE, OR SELF- CONTAINED FUNCTION GENERATOR

6.8 PRIMARY ELEMENT SYMBOLS (Contd.)

LEVEL L

1

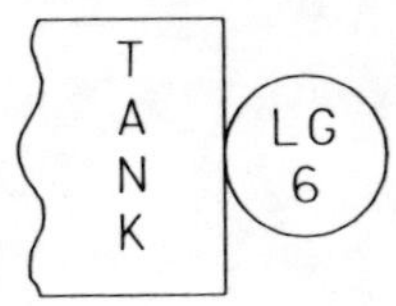

GAGE GLASS, INTEGRALLY MOUNTED ON TANK

2

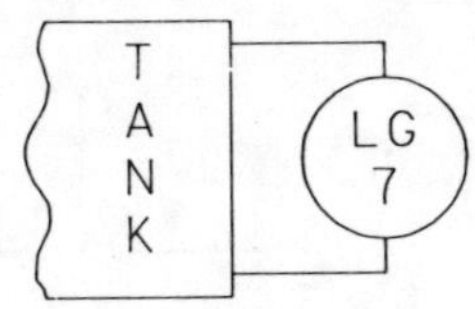

GAGE GLASS, EXTERNALLY CONNECTED

3

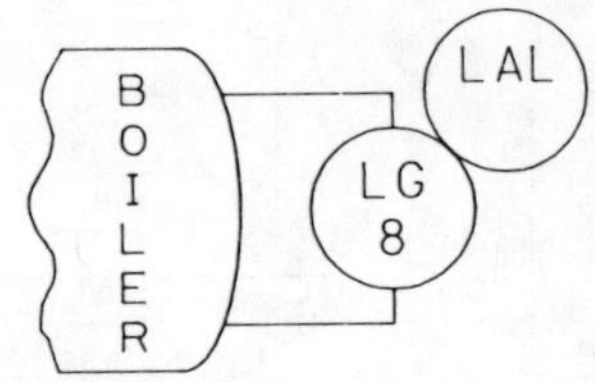

WATER COLUMN WITH INTEGRAL GAGE GLASS AND ALARM WHISTLE

4

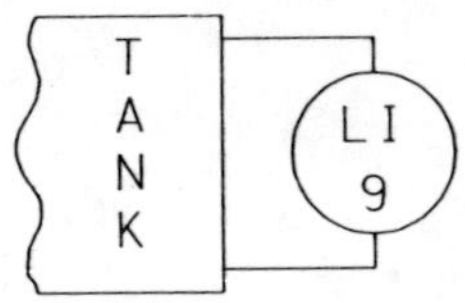

LEVEL INDICATOR, WITH TWO CONNECTIONS

5

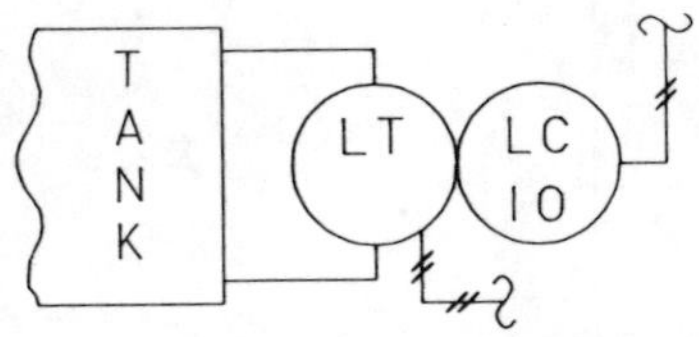

DUPLEX LEVEL TRANSMITTER-CONTROLLER, WITH TWO CONNECTIONS

6

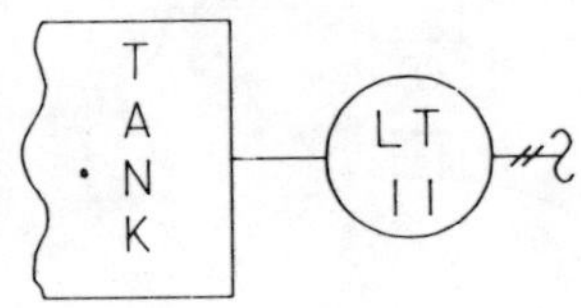

LEVEL TRANSMITTER, WITH ONE CONNECTION

7

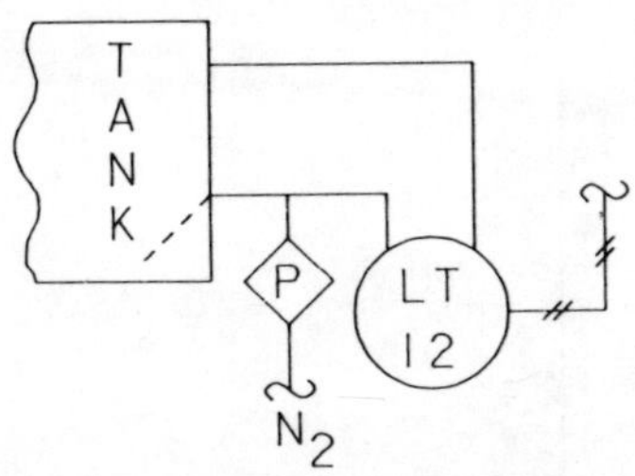

LEVEL TRANSMITTER, DIFFERENTIAL-PRESSURE TYPE, EXTERNALLY CONNECTED, WITH DIP TUBE

8

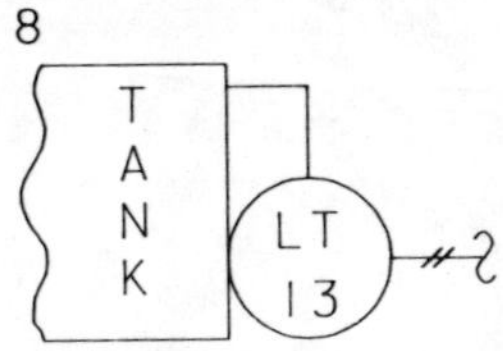

LEVEL TRANSMITTER, DIFFERENTIAL-PRESSURE TYPE, MOUNTED ON TANK

9

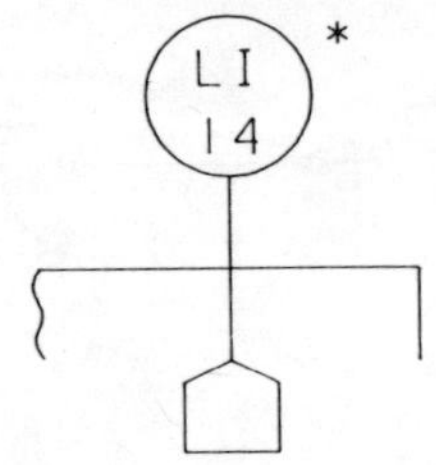

LEVEL INDICATOR, FLOAT TYPE

10

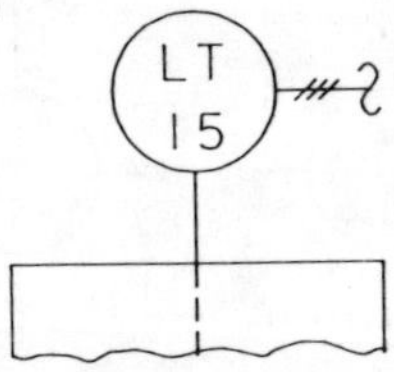

CAPACITANCE OR DIELECTRIC TYPE LEVEL ELEMENT CONNECTED TO LEVEL TRANSMITTER (TAG LEVEL ELEMENT LE-15)

11

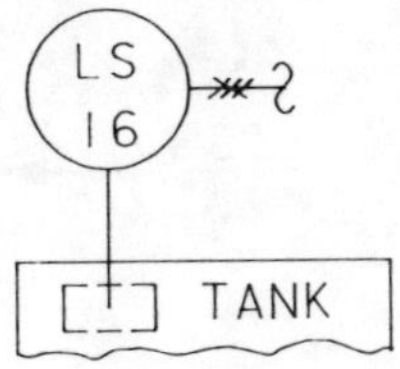

LEVEL SWITCH, PADDLE WHEEL OR LEVER TYPE, TO MEASURE LEVEL OF SOLIDS

12

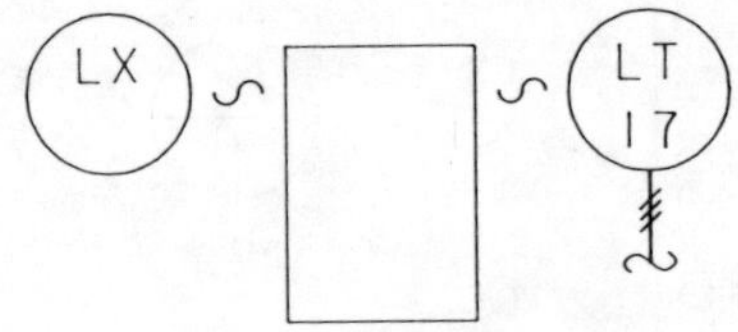

RADIOACTIVE- OR SONIC-TYPE LEVEL TRANSMITTER WITH INTEGRAL SENSOR

* Notations such as "mounted at grade" may be added.

6.8 PRIMARY ELEMENT SYMBOLS (Contd.)

L — LEVEL (Contd.)

13

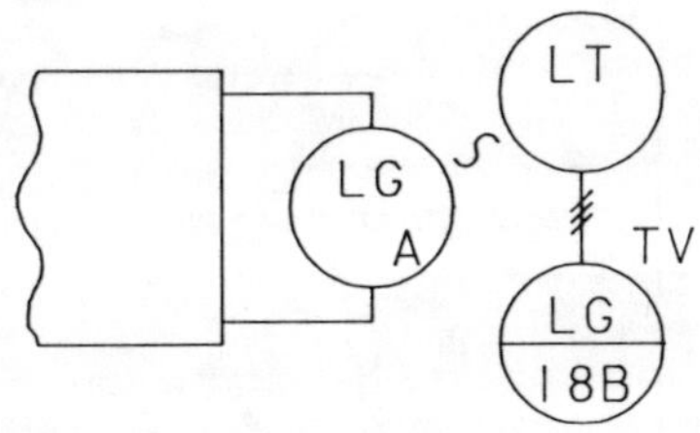

REMOTE VIEWING OF GAGE GLASS BY USE OF TELEVISION

14

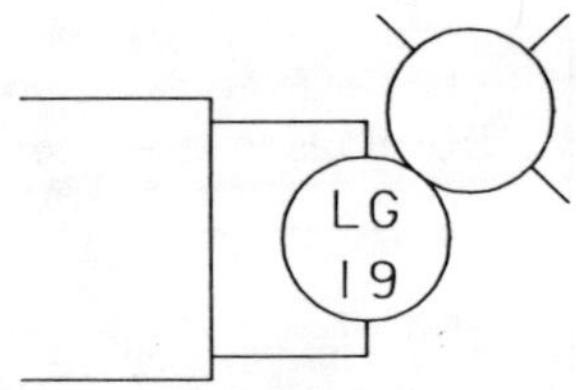

LEVEL GLASS WITH ILLUMINATOR

15

M — USER'S CHOICE

1

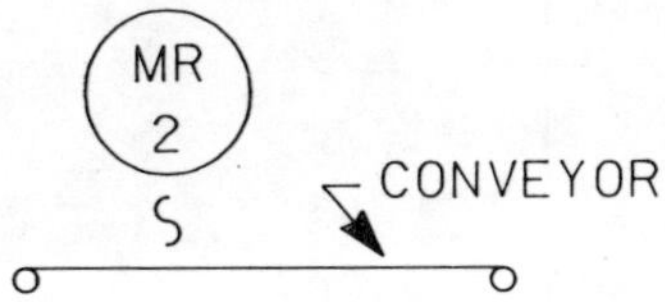

MOISTURE RECORDER (IF THERE IS A SEPARATE PRIMARY ELEMENT, IT SHOULD BE TAGGED ME-2

2

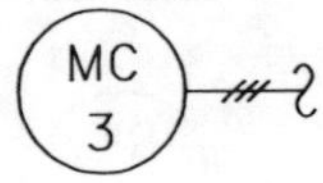

SELF-CONTAINED HUMIDITY CONTROLLER IN ROOM

USE OF LETTER M TO BE DEFINED IN USER'S LEGEND

P — PRESSURE OR VACUUM

1

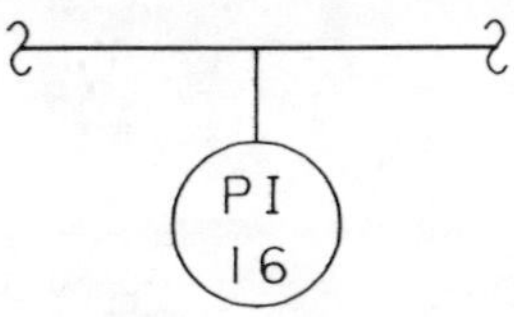

PRESSURE INDICATOR, DIRECT-CONNECTED

2

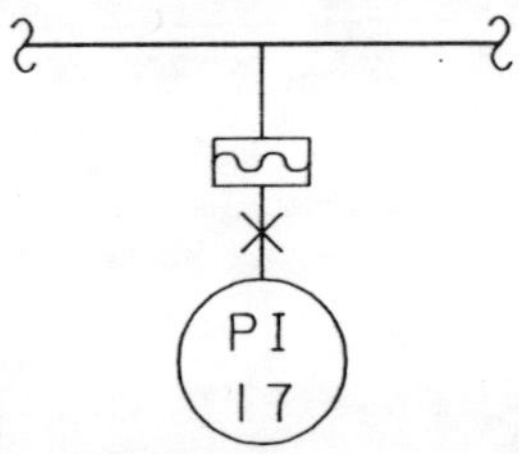

WITH PRESSURE LEAD LINE

3

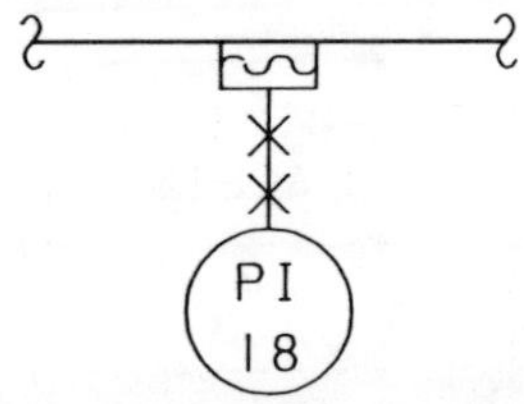

LINE-MOUNTED

PRESSURE INDICATOR CONNECTED TO DIAPHRAGM SEAL WITH FILLED SYSTEM

4

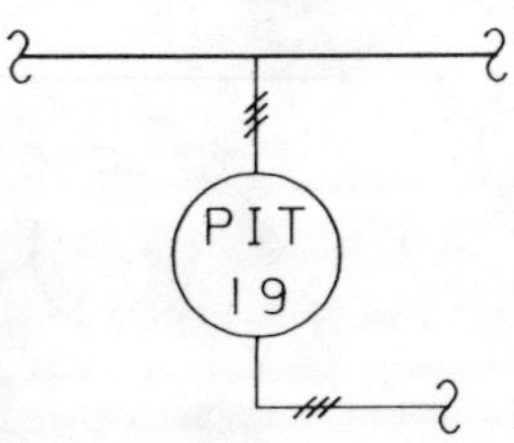

PRESSURE ELEMENT, STRAIN-GAGE TYPE, CONNECTED TO PRESSURE INDICATING TRANSMITTER (TAG STRAIN GAGE PE-19)

5

6

6.8 PRIMARY ELEMENT SYMBOLS (Contd.)

	1	2	3
Q QUANTITY	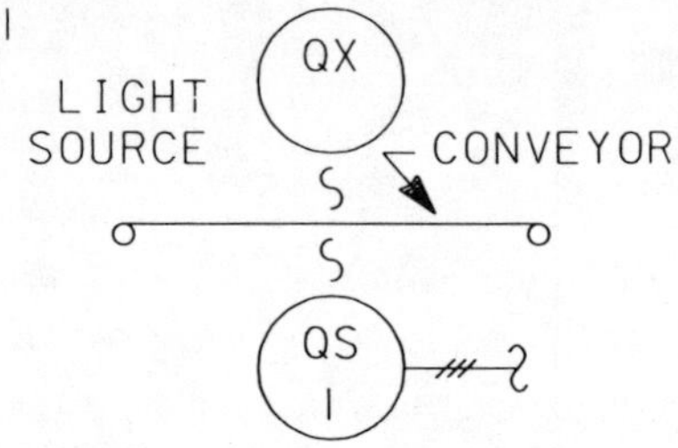 COUNTING SWITCH, PHOTO-ELECTRIC TYPE, WITH SWITCH ACTION FOR EACH EVENT	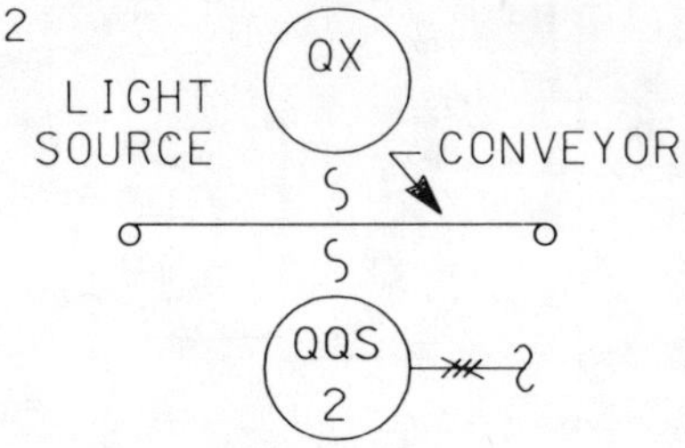 COUNTING SWITCH, PHOTO-ELECTRIC TYPE, WITH SWITCH ACTION BASED ON CUMULATIVE TOTAL	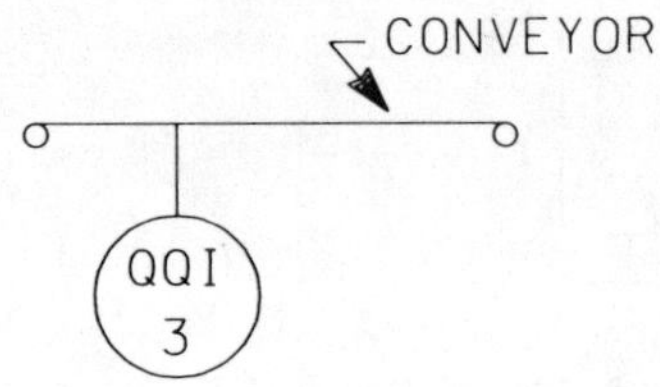 INDICATING COUNTER, MECHANICAL TYPE
R RADIATION	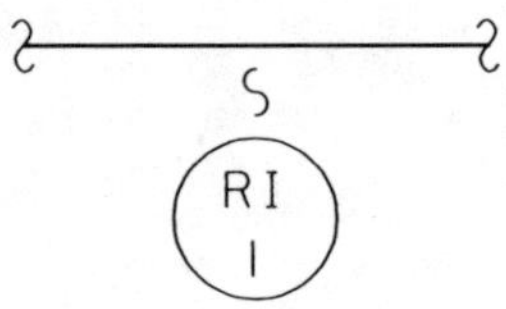 RADIATION INDICATOR	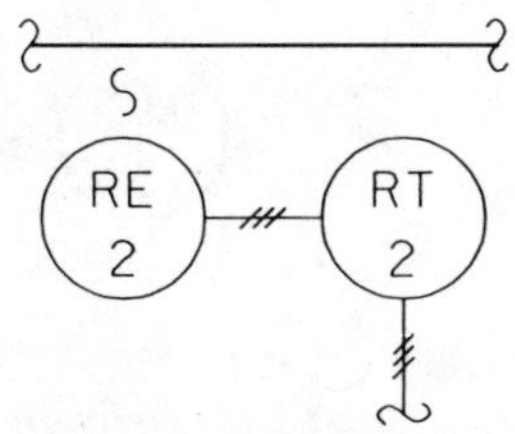 RADIATION MEASURING ELEMENT AND TRANSMITTER	
S SPEED OR FREQUENCY	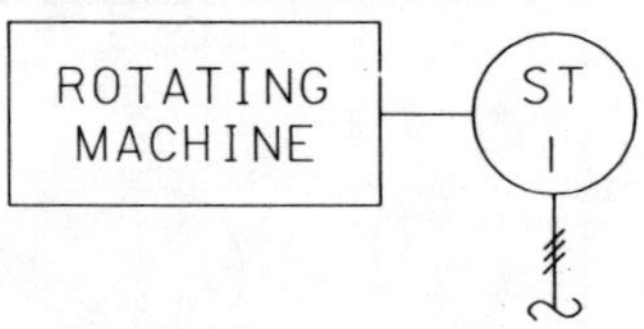 SPEED TRANSMITTER		
T TEMPERATURE	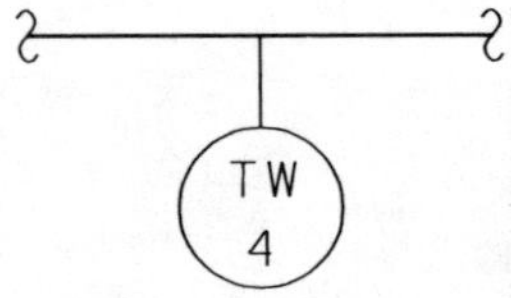 TEMPERATURE CONNECTION WITH WELL	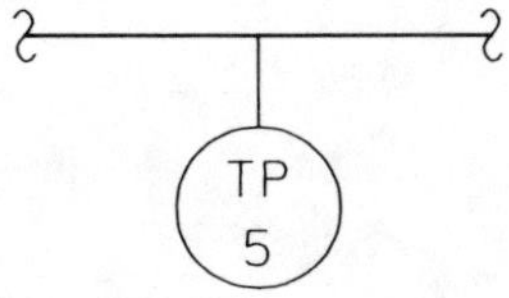 TEMPERATURE TEST CONNECTION WITHOUT WELL	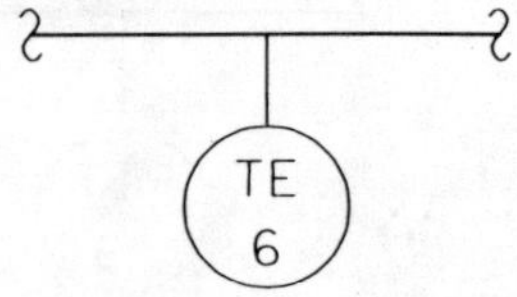 TEMPERATURE ELEMENT WITHOUT WELL (ELEMENT NOT CONNECTED TO SECONDARY INSTRUMENT)

6.8 PRIMARY ELEMENT SYMBOLS (Contd.)

T

TEMPERATURE (Contd.)

4 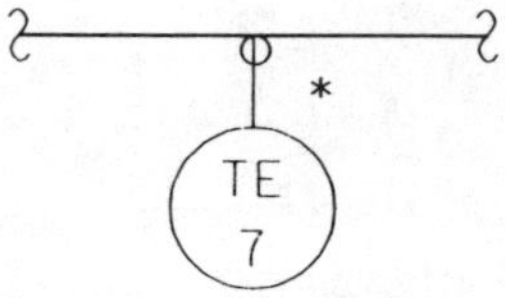 TEMPERATURE ELEMENT WITH WELL (ELEMENT NOT CONNECTED TO SECONDARY INSTRUMENT)	5 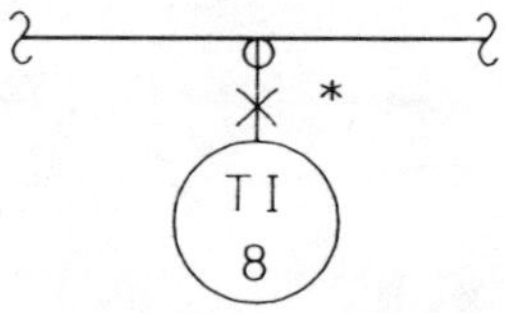 FILLED-SYSTEM-TYPE TEMPERATURE INDICATOR WITH WELL	6 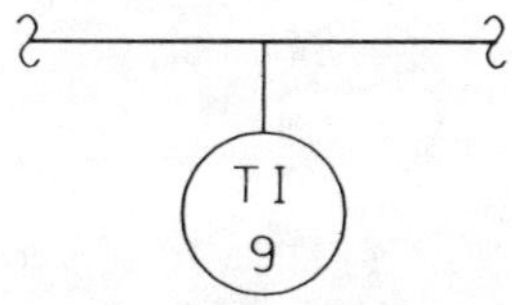 BIMETALLIC-TYPE THERMOMETER, GLASS THERMOMETER, OR OTHER LOCAL UNCLASSIFIED TEMPERATURE INDICATOR
7 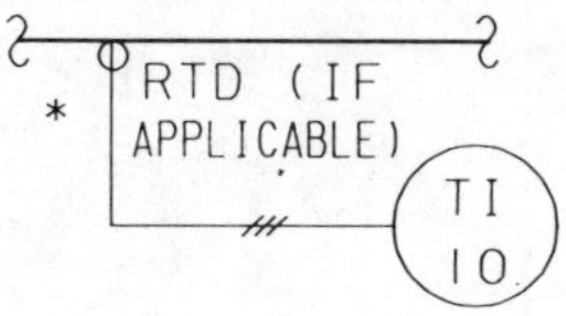 THERMOCOUPLE, RESISTANCE BULB (RTD) OR THERMISTOR (TH) CONNECTED TO TEMPERATURE INDICATOR (TAG ELEMENT TE-10)	8 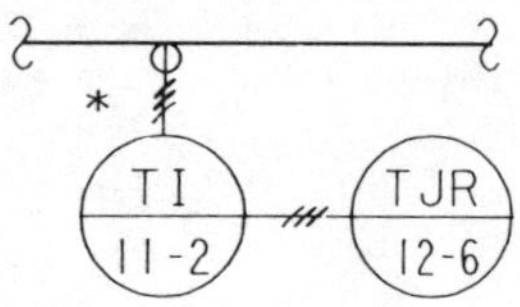 THERMOCOUPLE CONNECTED TO MULTIPOINT INDICATOR RECORDING ON MULTIPOINT SCANNING RECORDER (TAG ELEMENT TE-11-2)	9 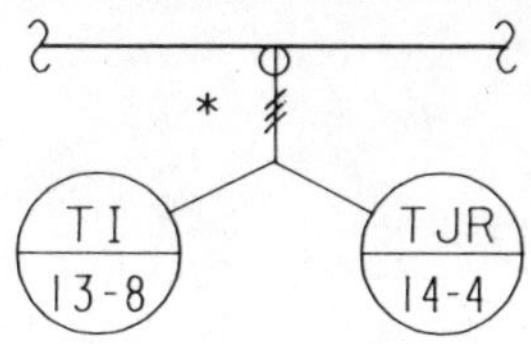 THERMOCOUPLE PARALLEL-WIRED TO MULTIPOINT INDICATOR AND MULTIPOINT SCANNING RECORDER (TAG ELEMENT TE-13-8/14-4)
10 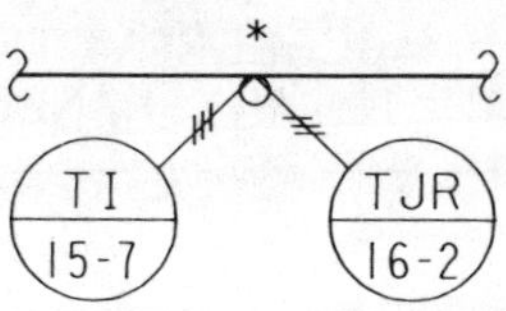 DUAL OR DUPLEX THERMOCOUPLE CONNECTED TO MULTIPOINT INDICATOR AND MULTIPOINT SCANNING RECORDER (TAG ELEMENT TE-15-7/16-2)	11 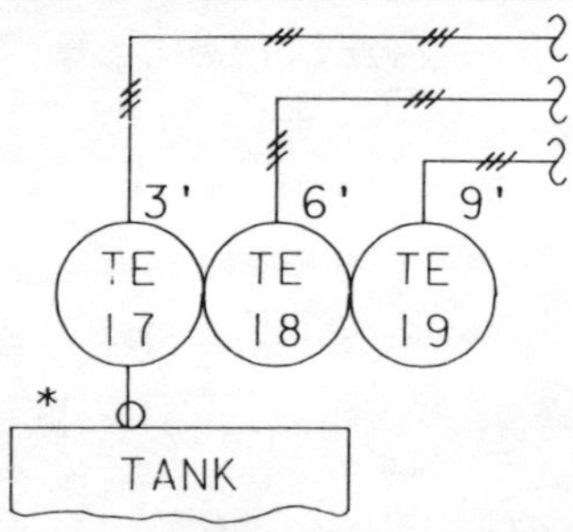 MULTI-ELEMENT THERMOCOUPLE FOR DIFFERENT ELEVATIONS, WITH WELL IN TANK	12 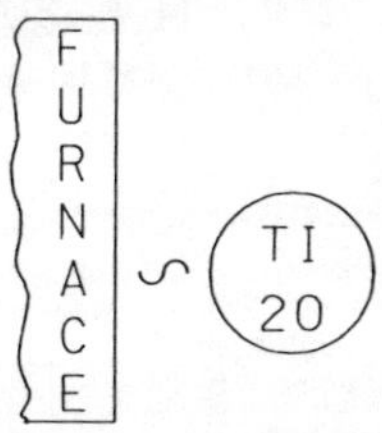 THERMAL-RADIATION TYPE TEMPERATURE INDICATOR, SELF-CONTAINED
13 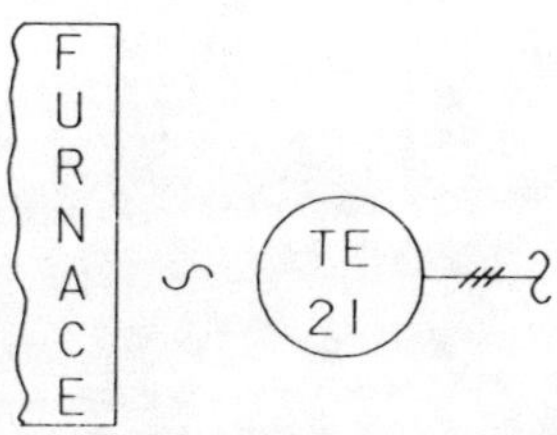 THERMAL-RADIATION-TYPE TEMPERATURE ELEMENT	14 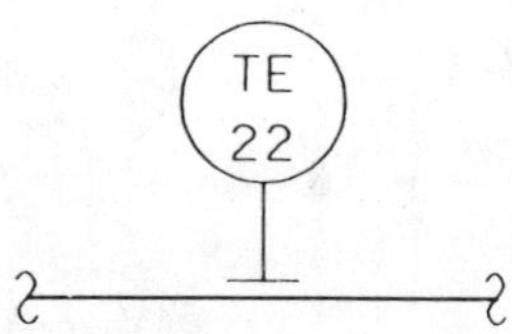 SURFACE-MOUNTED TEMPERATURE SENSOR	15 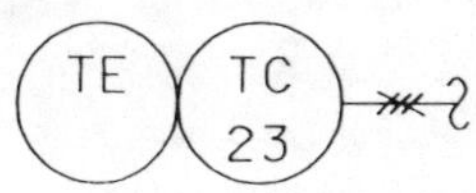 THERMOSTAT SENSING AMBIENT TEMPERATURE

* Use of the thermowell symbol is optional. However, use or omission of the symbol should be consistent throughout a project.

6.8 PRIMARY ELEMENT SYMBOLS (Contd.)

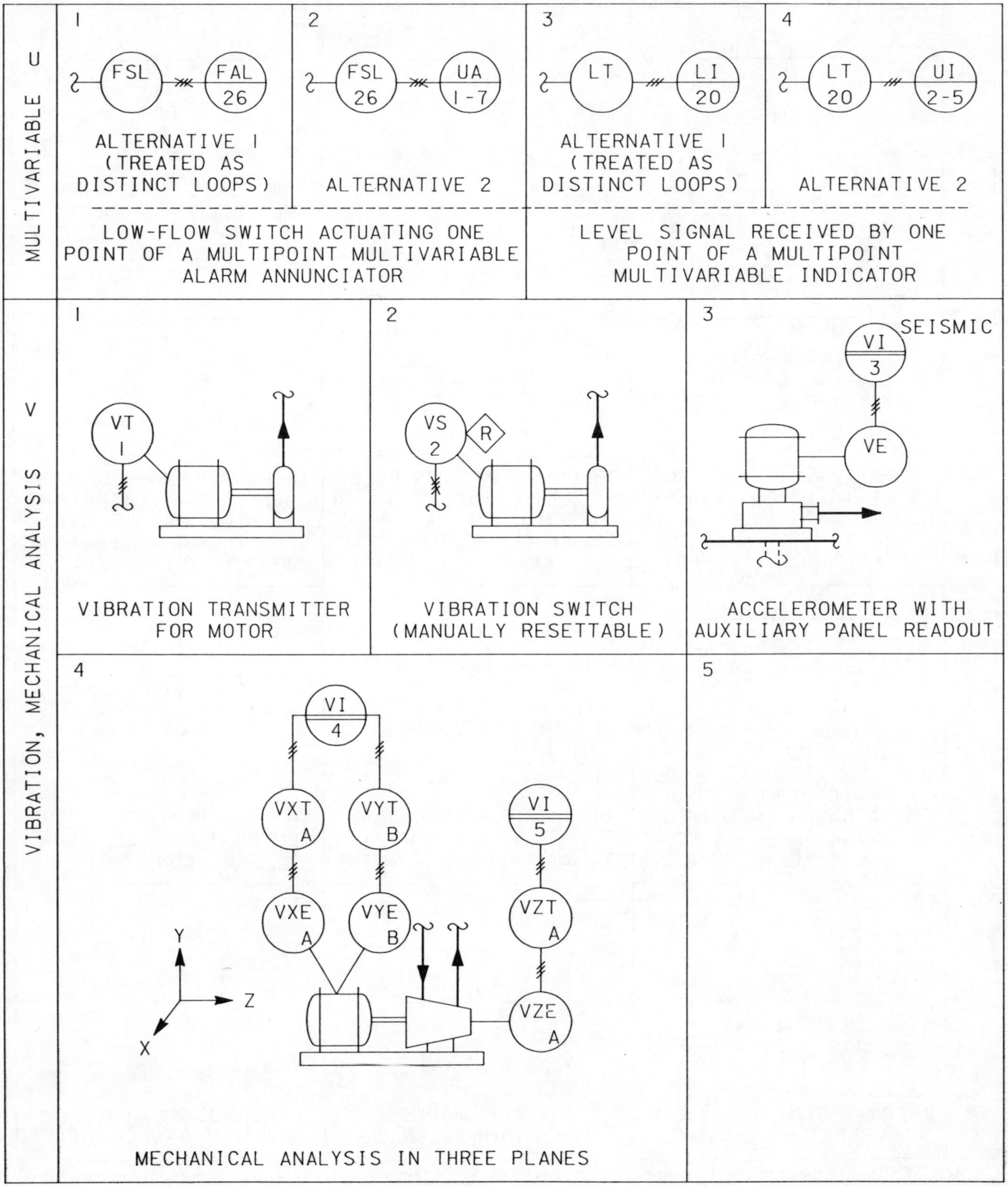

6.8 PRIMARY ELEMENT SYMBOLS (Contd.)

W — WEIGHT OR FORCE

1

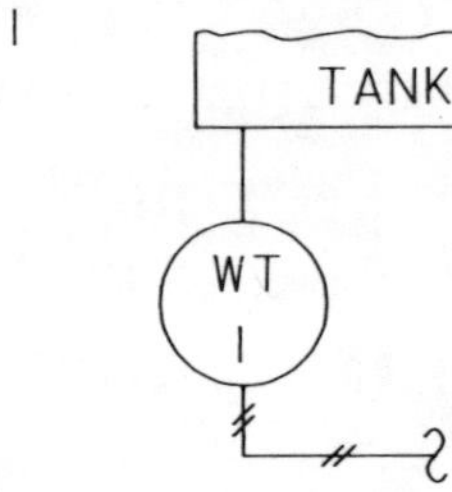

WEIGHT TRANSMITTER, DIRECT-CONNECTED

2

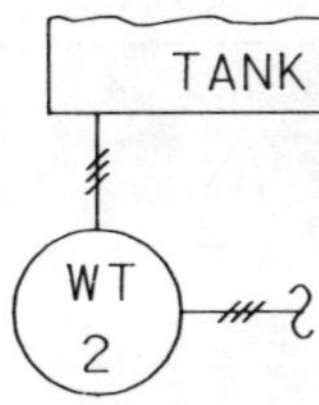

STRAIN GAGE CONNECTED TO SEPARATE WEIGHT TRANSMITTER (TAG STRAIN GAGE WE-2)

3

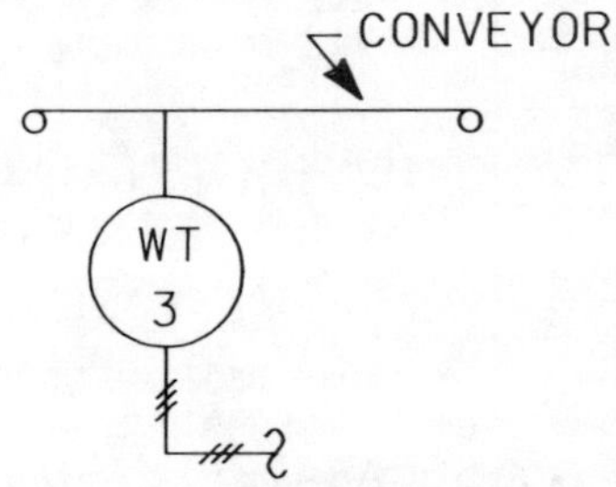

WEIGH-BELT SCALE TRANSMITTER

Z — POSITION, DIMENSION

1

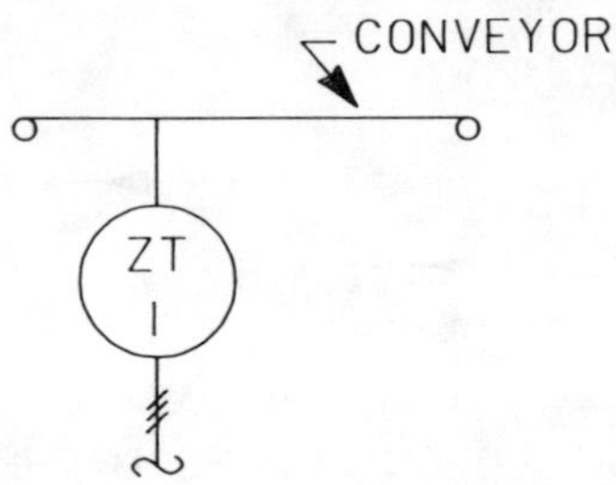

ROLL-THICKNESS TRANSMITTER

2

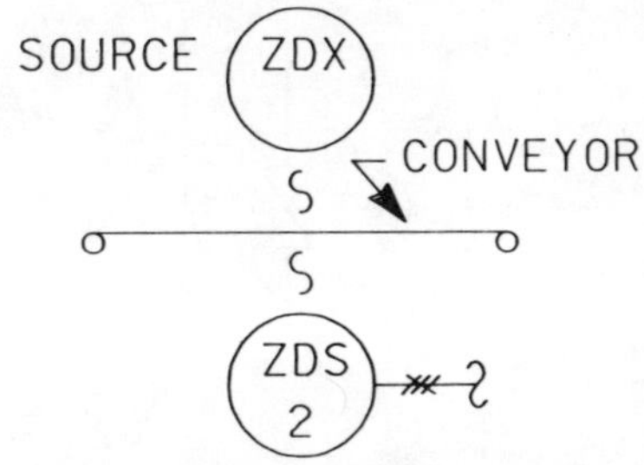

THICKNESS SWITCH, RADIOACTIVE TYPE

3

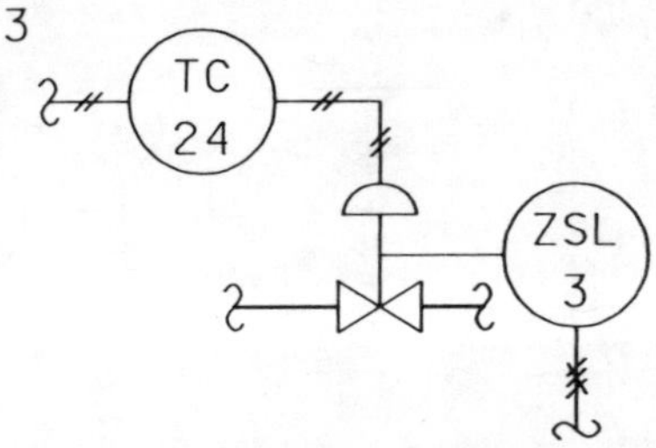

LIMIT SWITCH THAT IS ACTUATED WHEN VALVE IS CLOSED TO A PRE-DETERMINED POSITION

4

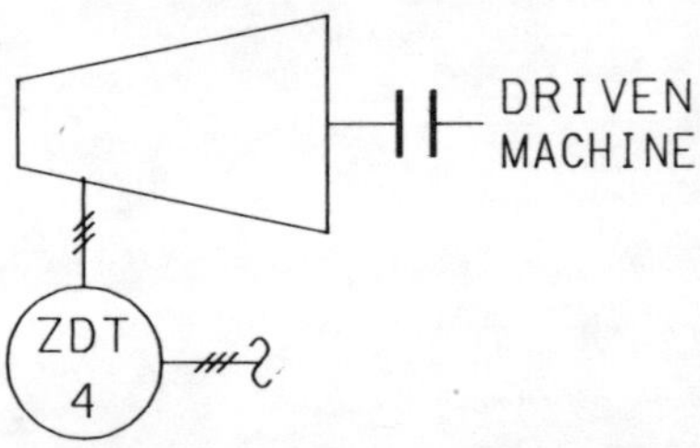

TURBINE SHELL/ROTOR DIFFERENTIAL-EXPANSION TRANSMITTER (TAG PRIMARY ELEMENT ZDE-4)

5

6

6.9 EXAMPLES-FUNCTIONS

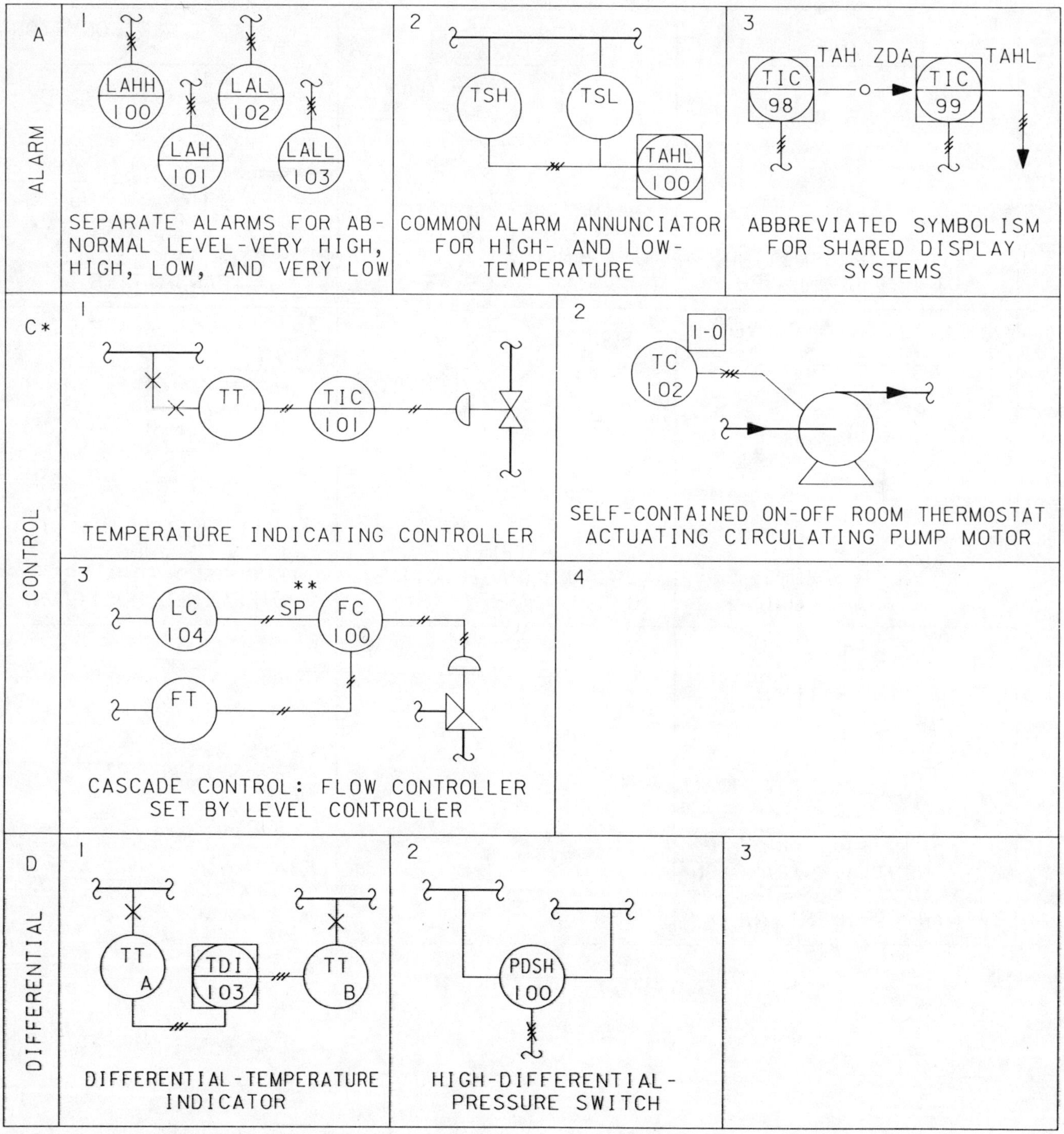

* It is expected that control modes will not be designated on a diagram. However, designations may be used outside the controller symbol, if desired, in combinations such as [%], [∫], [I-0].

** A controller is understood to have integral manual set-point adjustment unless means of remote adjustment is indicated. The remote set-point designation is SP.

6.9 EXAMPLES-FUNCTIONS (Contd.)

F — RATIO

1

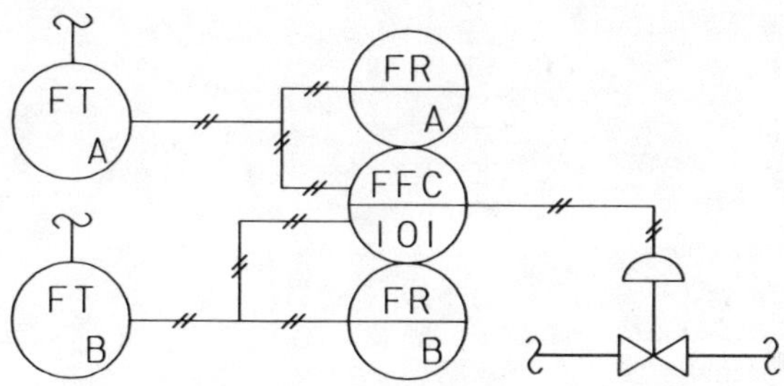

FLOW-RATIO CONTROLLER WITH TWO PENS TO RECORD FLOW

2

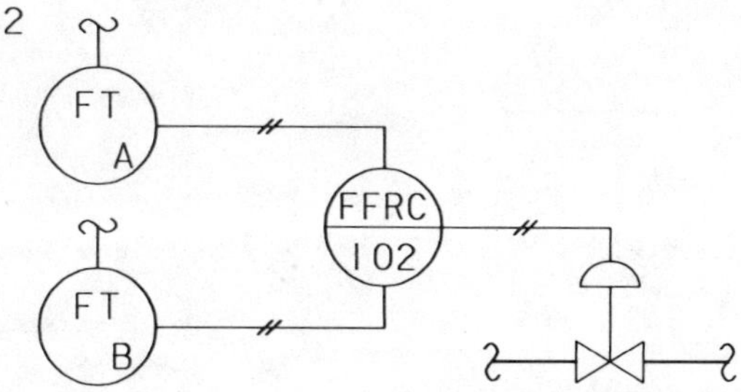

FLOW-RATIO CONTROLLER WITH ONE PEN TO RECORD FLOW-RATIO

3

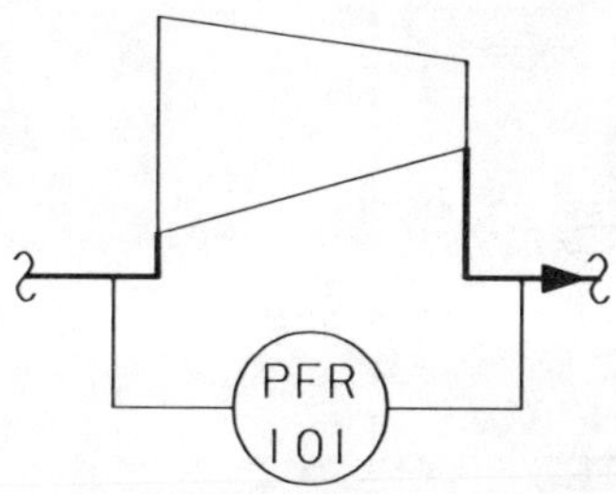

DIRECT-CONNECTED COMPRESSION-RATIO RECORDER

4

G — VIEWING DEVICE, GLASS

1

SIGHT GLASS FOR INTERNAL VIEWING

2

3

H — HAND

1

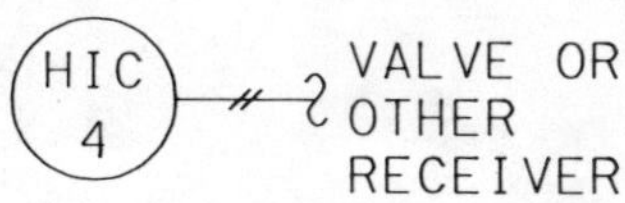

MANUAL LOADING STATION WITH OUTPUT GAGE

2

HAND-ACTUATED ELECTRIC SWITCH, MOMENTARY

3

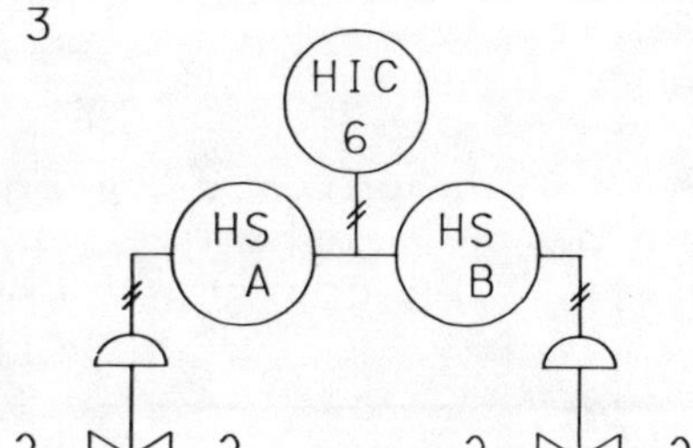

MANUAL LOADING STATION WITH HAND ACTUATED SWITCHES

6.9 EXAMPLES-FUNCTIONS (Contd.)

I — INDICATE

1

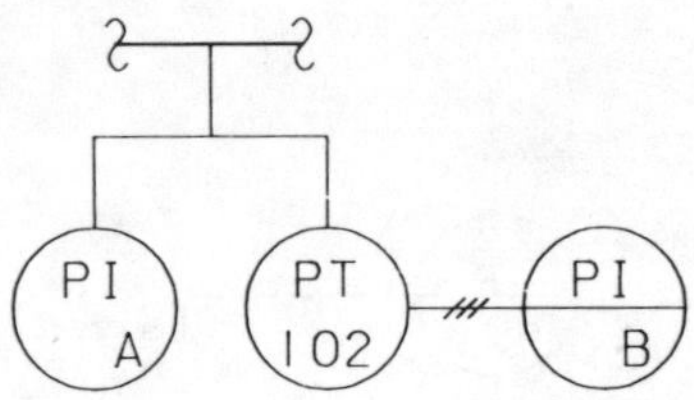

LOCAL PRESSURE INDICATOR AND PRESSURE TRANSMITTER WITH COMMON TAP AND PANEL-MOUNTED PRESSURE INDICATOR

2

3

J — SCAN

1

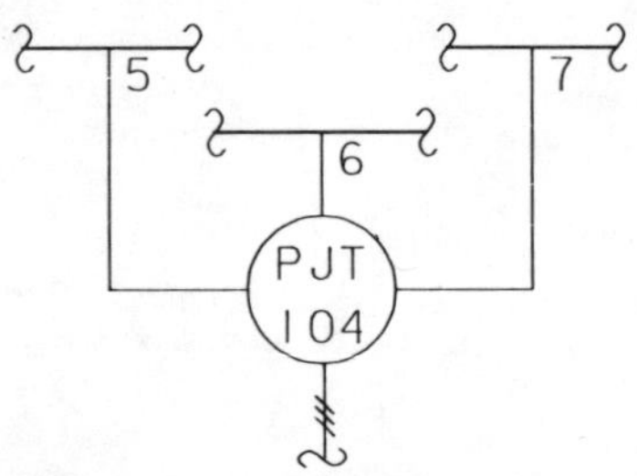

PRESSURE-SCANNING TRANSMITTER CONNECTED TO PROCESS POINTS 5,6,7

2

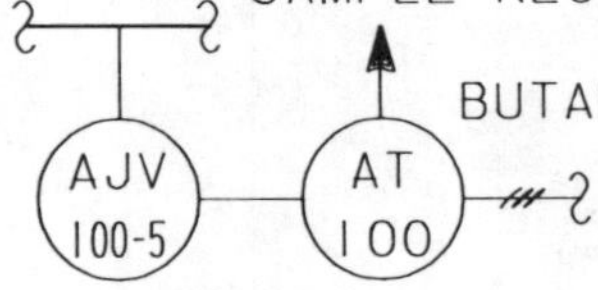

SAMPLE LINE CONNECTED TO BUTANE-CONCENTRATION TRANSMITTER THROUGH EXTERNAL SAMPLE-SCANNING VALVE

3

K — CONTROL STATION

1

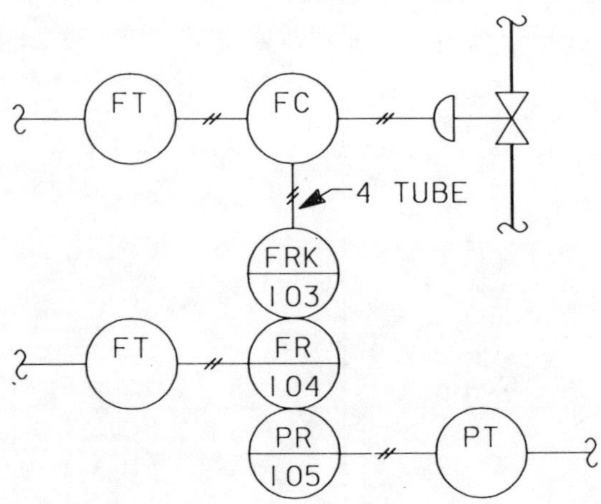

RECORDING FLOW CONTROL STATION PANEL MOUNTED, WITH ADDITIONAL FLOW AND PRESSURE PENS, AND LOCAL CONTROLLER

2

6.9 EXAMPLES-FUNCTIONS (Contd.)

L — LIGHT OR LOW

1

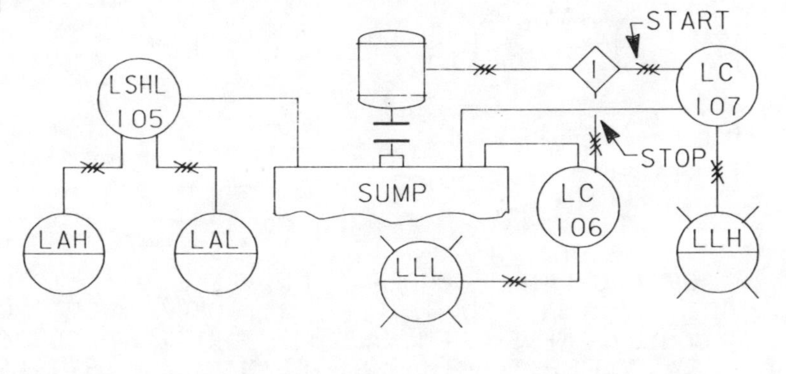

DIFFERENTIAL-GAP CONTROL OF SUMP LEVEL THROUGH STARTING AND STOPPING SUMP PUMP BY LC-107 AND LC-106 THAT ALSO ACTUATE HIGH- AND LOW-LEVEL PILOT LIGHTS. HIGH- AND LOW-LEVEL ALARMS ARE ACTUATED BY LSHL-105

2

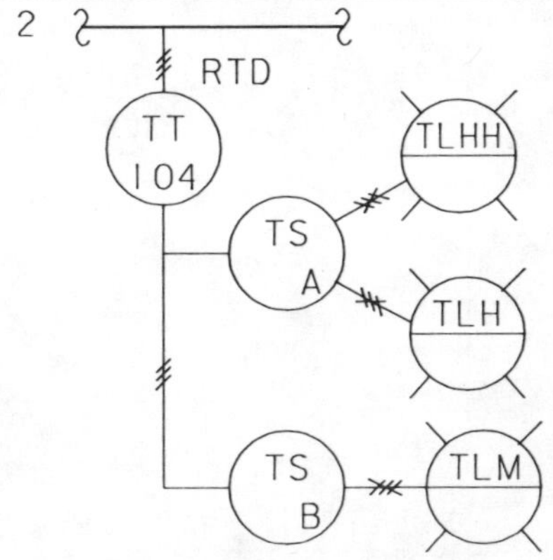

PILOT LIGHTS TO SIGNAL THAT TEMPERATURE HAS RISEN TO INTERMEDIATE, HIGH, AND VERY HIGH VALUES

P — POINT

1

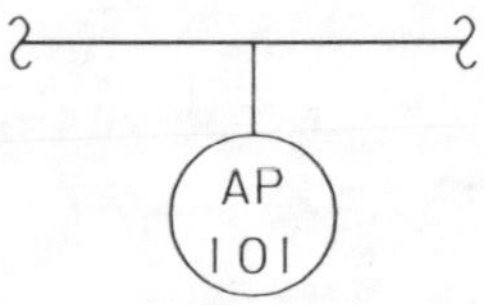

ANALYSIS TEST SAMPLE POINT

2

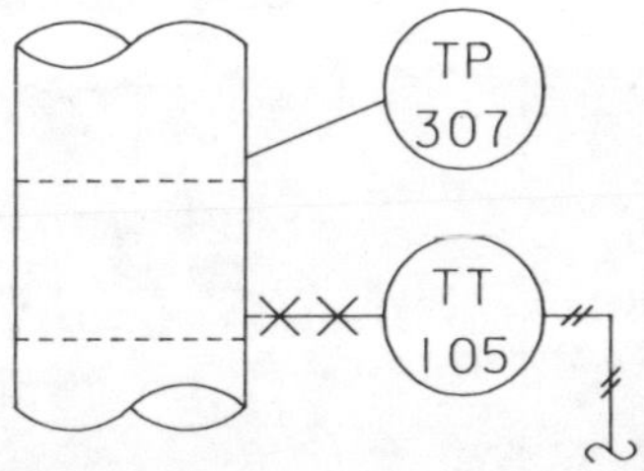

DISTILLATION COLUMN WITH CONNECTION FOR ALTERNATIVE LOCATION OF SENSOR

3

Q — INTEGRATE OR TOTALIZE

1

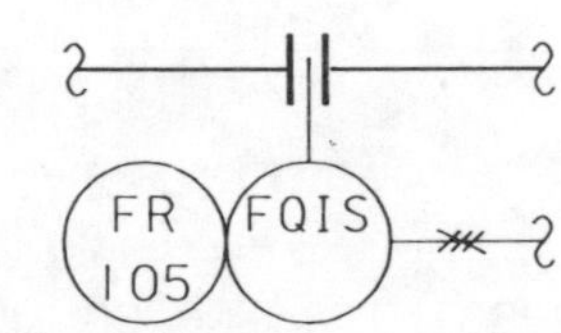

DIFFERENTIAL-PRESSURE-TYPE FLOW METER WITH (1) RECORDING OF FLOW RATE, (2) INDICATION OF INTEGRATED FLOW, AND (3) SWITCH ACTUATED BY INTEGRATED FLOW.

2

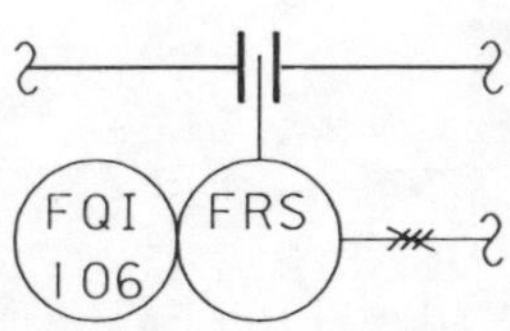

DIFFERENTIAL-PRESSURE-TYPE FLOW METER WITH (1) RECORDING OF FLOW RATE, (2) INDICATION OF INTEGRATED FLOW, AND (3) SWITCH ACTUATED BY FLOW RATE.

3

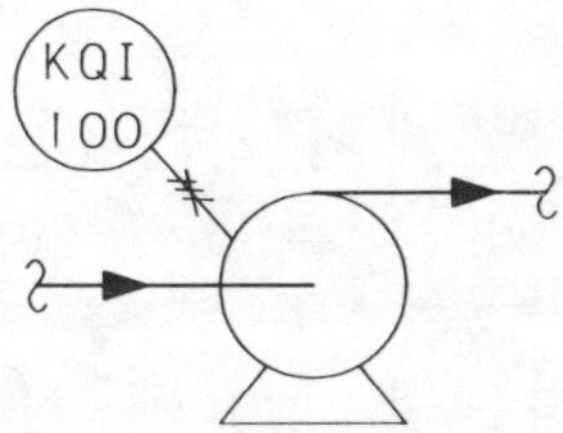

RUNNING-TIME TOTALIZER CONNECTED TO PUMP MOTOR

6.9 EXAMPLES-FUNCTIONS (Contd.)

R — RECORD OR PRINT

1

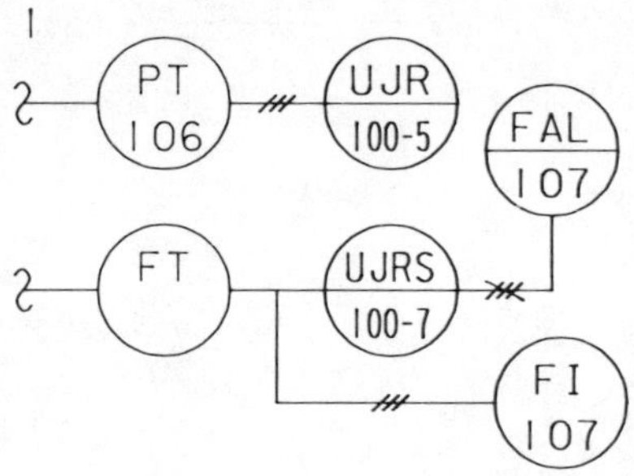

MULTIPOINT MULTI-VARIABLE DATA LOGGER, UJR-100, WITH SWITCH FOR POINT 7

2

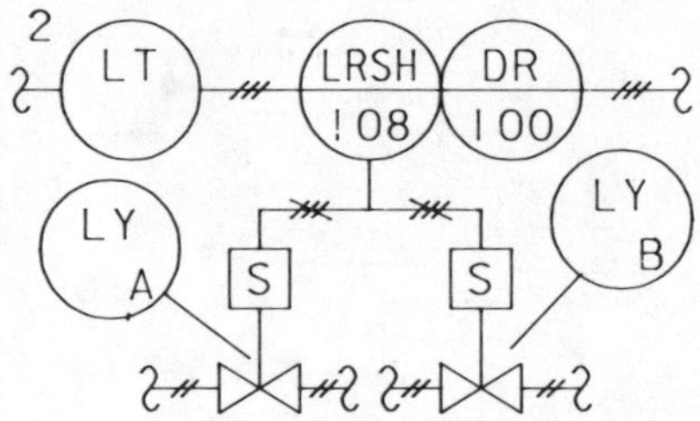

LEVEL SIGNAL TO ONE PEN OF TWO-PEN RECORDER. BOTH PENS RECORD CONTINUOUSLY. LEVEL PEN HAS SWITCH THAT ACTUATES TWO SOLENOID VALVES ON HIGH LEVEL

3

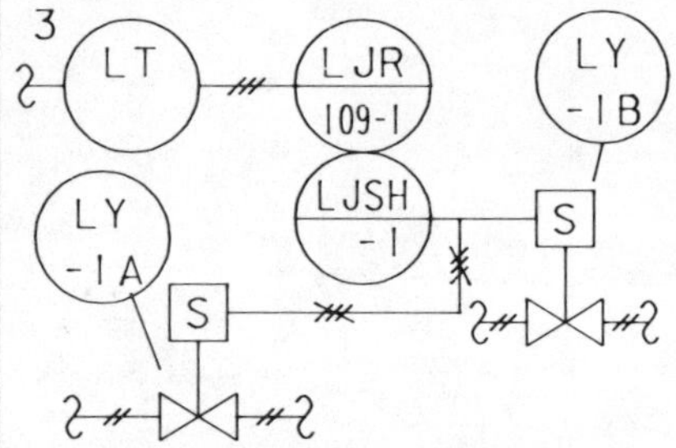

LEVEL SIGNAL TO POINT 1 OF A MULTI-POINT LEVEL RECORDER. POINT 1 HAS SWITCH THAT ACTUATES TWO SOLENOID VALVES ON HIGH LEVEL

S — SWITCH

1

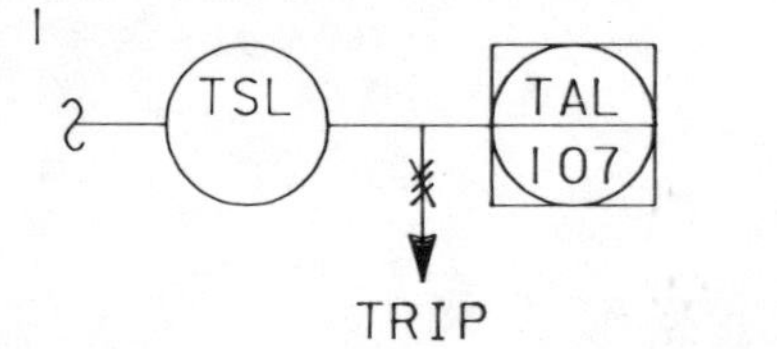

LOW TEMPERATURE SWITCH WITH ONE OUTPUT FOR ALARM AND TRIP

2

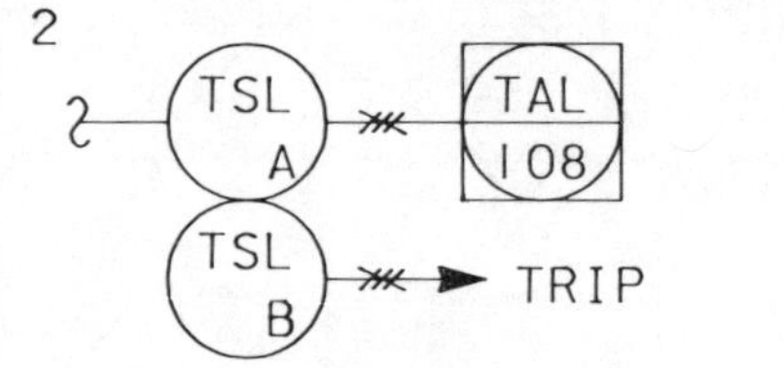

ALTERNATIVE 1

3

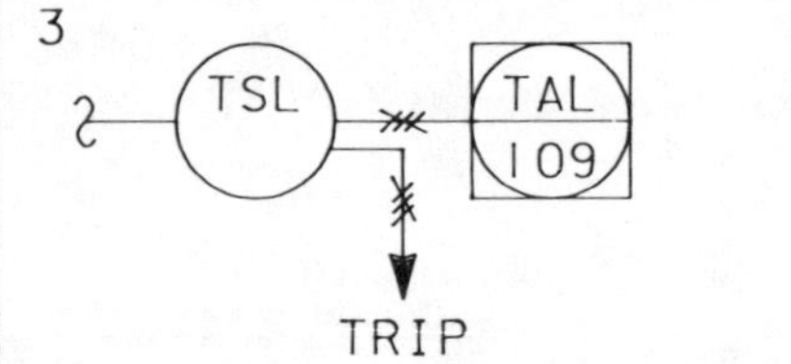

ALTERNATIVE 2

LOW-TEMPERATURE SWITCH WITH TWO OUTPUTS FOR ALARM AND TRIP

4

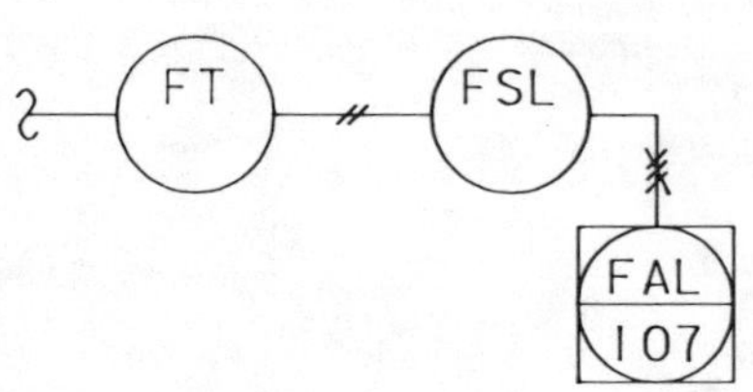

RECEIVER-TYPE FLOW SWITCH ACTUATING LOW-FLOW ALARM

5

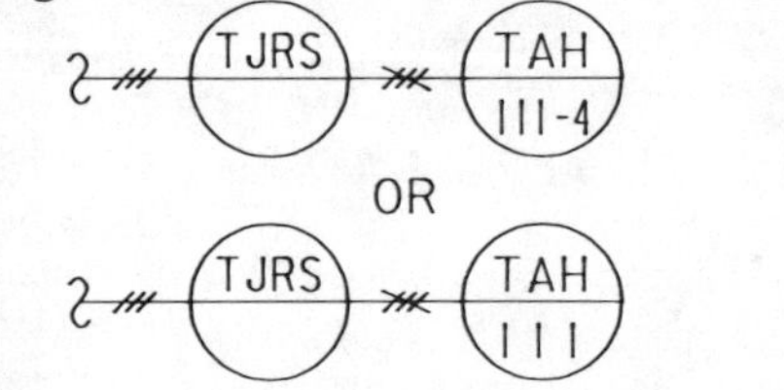

ALTERNATIVE 1

6

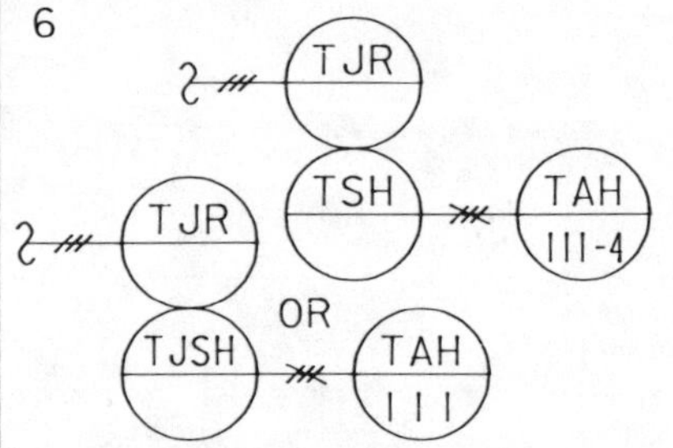

ALTERNATIVE 2

MULTIPOINT TEMPERATURE RECORDER WITH INTEGRAL HIGH-TEMPERATURE SWITCH FOR POINT 4, OR COMMON ALARM

7

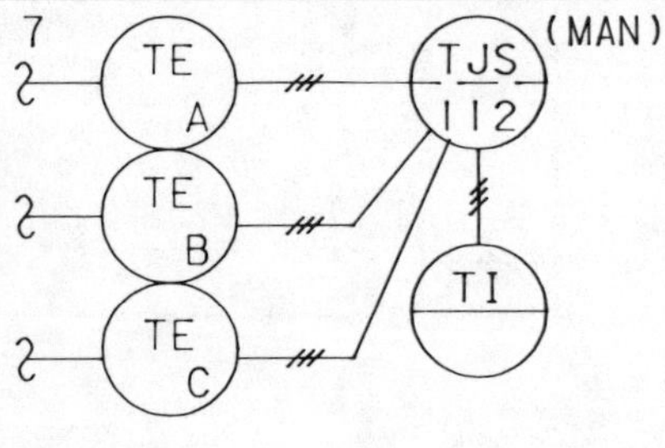

ONE-POINT TEMPERATURE INDICATOR WITH SEPARATE MANUAL SELECTOR SWITCH

8

9

6.9 EXAMPLES-FUNCTIONS (Contd.)

T — TRANSMIT

1

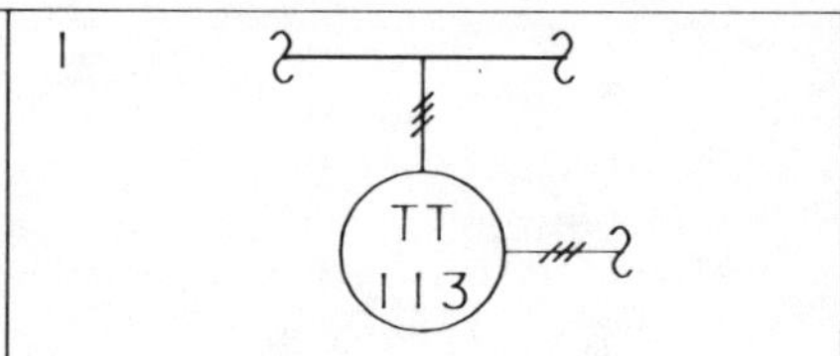

TRANSMITTER WITH THERMO-COUPLE INPUT AND ELECTRIC OUTPUT*

2

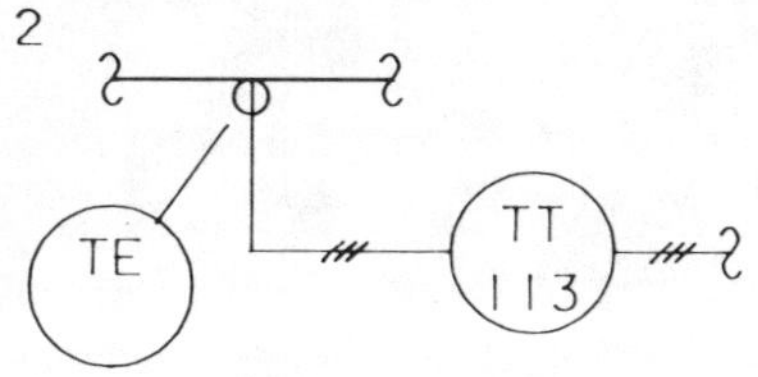

ALTERNATIVE TO 1

3

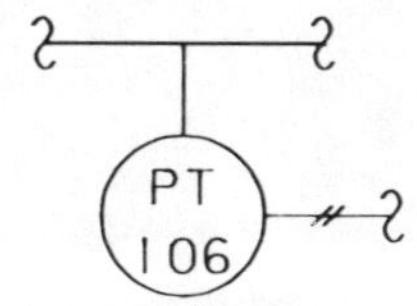

TRANSMITTER WITH PNEUMATIC OUTPUT

U — MULTIFUNCTION

1

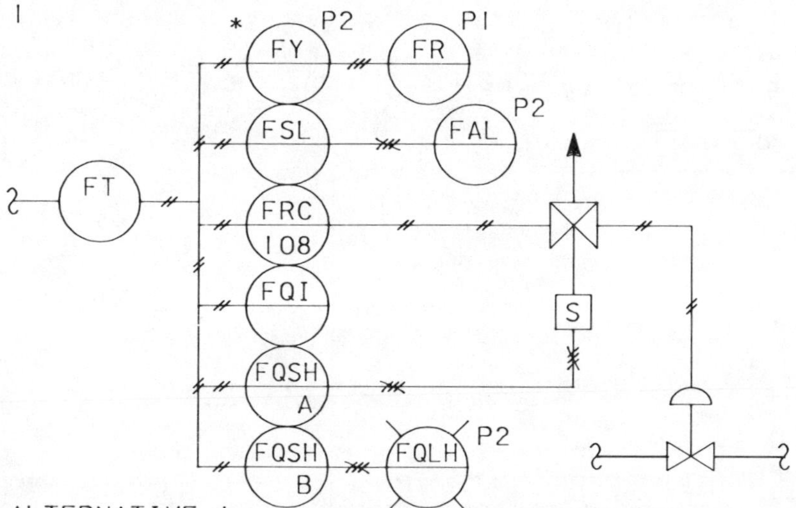

ALTERNATIVE 1

2

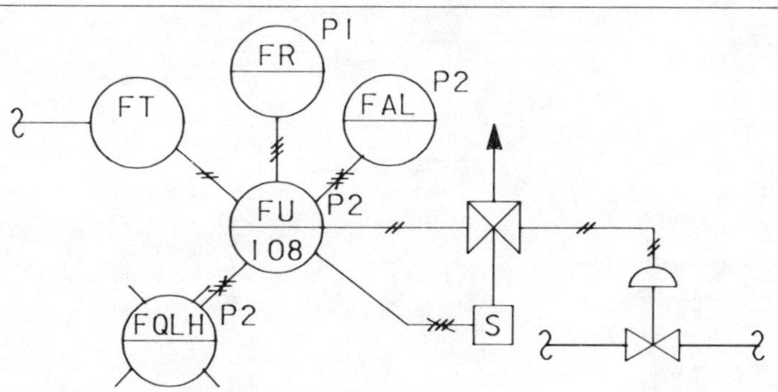

ALTERNATIVE 2

FLOW SIGNAL RECEIVED BY PANEL P2 INSTRUMENT THAT INDICATES AND RECORDS FLOW, HAS SWITCH TO ACTUATE LOW-FLOW ALARM, HAS A CONTROLLER TO ACTUATE A VALVE, RELAYS THE MEASUREMENT TO A RECORDER ON PANEL P1, INTEGRATES THE FLOW, INDICATES THE INTEGRATED FLOW, HAS A SWITCH TO ACTUATE A SOLENOID VALVE AFTER A GIVEN INTEGRATED FLOW HAS PASSED, AND HAS ANOTHER SWITCH TO ACTUATE A PILOT LIGHT AFTER ANOTHER INTEGRATED FLOW HAS PASSED

V — VALVE, DAMPER, OR LOUVER

1

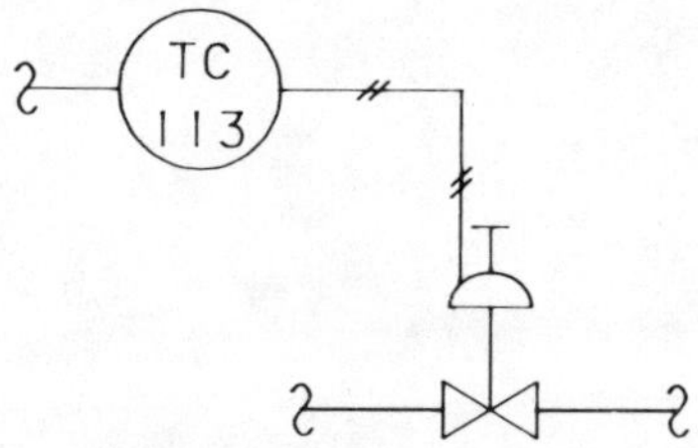

CONTROL VALVE WITH TOP-MOUNTED HANDJACK

2

3

* See definition of converter versus transmitter.

6.9 EXAMPLES-FUNCTIONS (Contd.)

Y

COMPUTE OR RELAY

1

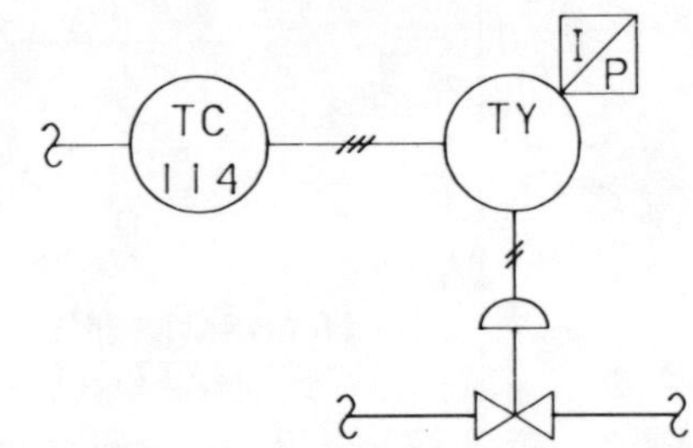

ELECTRIC CURRENT SIGNAL ACTUATING PNEUMATIC CONTROL VALVE WITH SEPARATELY MOUNTED ELECTRO-PNEUMATIC CONVERTER. THE CONVERTER SYMBOL MAY BE OMITTED

2

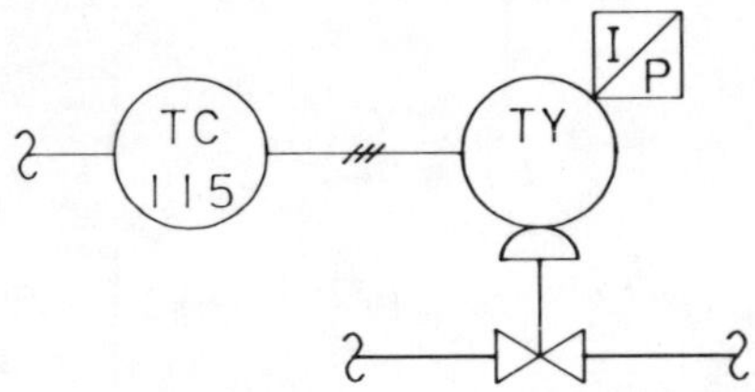

ELECTRIC CURRENT SIGNAL TO PNEUMATIC CONTROL VALVE FURNISHED WITH ATTACHED ELECTRO-PNEUMATIC SIGNAL CONVERTER. THE CONVERTER SYMBOL MAY BE OMITTED

3

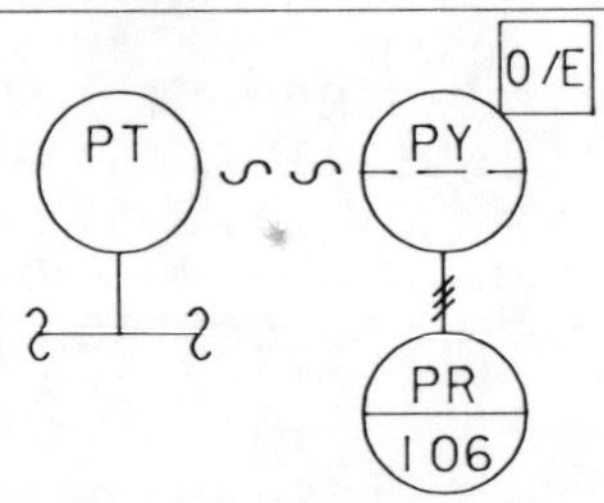

PRESSURE TRANSMITTER WITH RADIO (OR LASER) OUTPUT, RADIO- (OR LASER-) TO-VOLTAGE CONVERTER, AND PRESSURE RECORDER

4

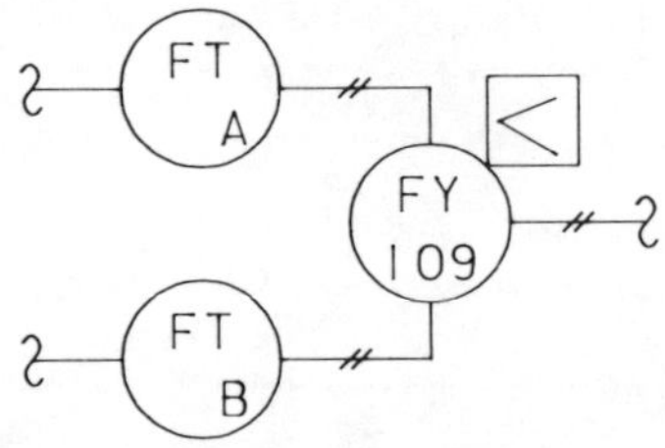

SELECTOR RELAY WHOSE OUTPUT REPRESENTS LOWER FLOW OF FT-109A OR FT-109B

5

6

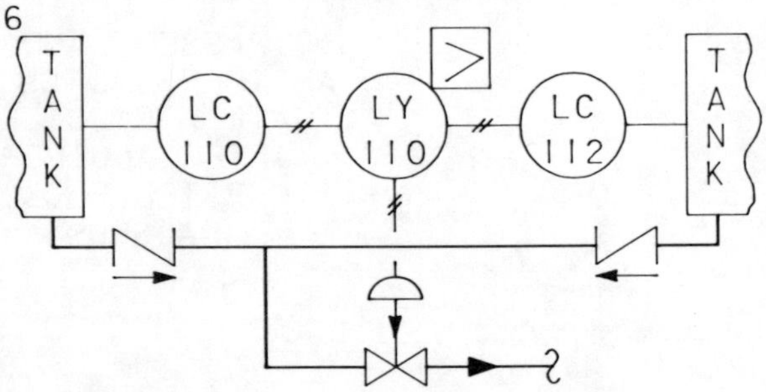

TANK OUTLET VALVE TO OPEN AS REQUIRED BY THE HIGHER OF TWO LEVELS

7

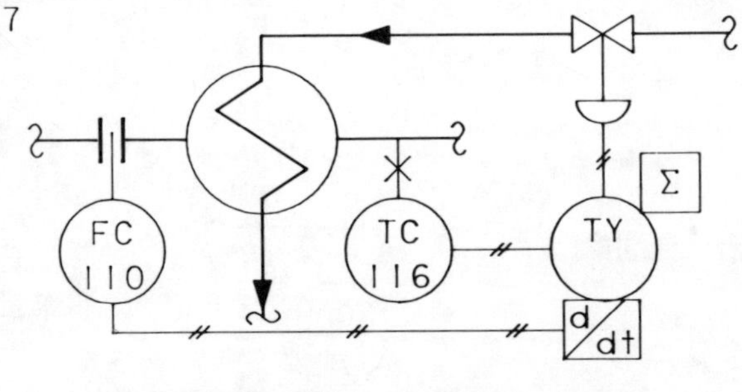

TEMPERATURE CONTROL WITH FLOW-ANTICIPATING RELAY

6.9 EXAMPLES-FUNCTIONS (Contd.)

<table>
<tr>
<td>Y
COMPUTE OR RELAY
(Contd.)</td>
<td colspan="2">8
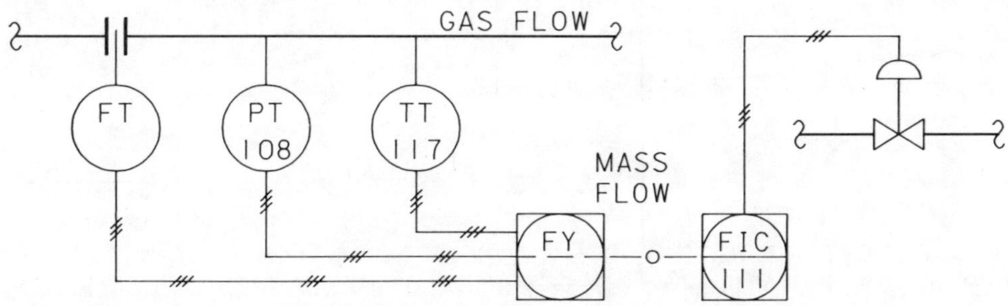

MASS-FLOW COMPUTATION PERFORMED BY MASS-FLOW COMPUTING RELAY FY-111 HAVING LINEAR INPUTS FOR PRESSURE AND TEMPERATURE AND A NON-LINEAR INPUT FOR FLOW</td>
</tr>
<tr>
<td>Z
DRIVE, ACTUATE, OR UNCLASSIFIED FINAL CONTROL ELEMENT</td>
<td>1
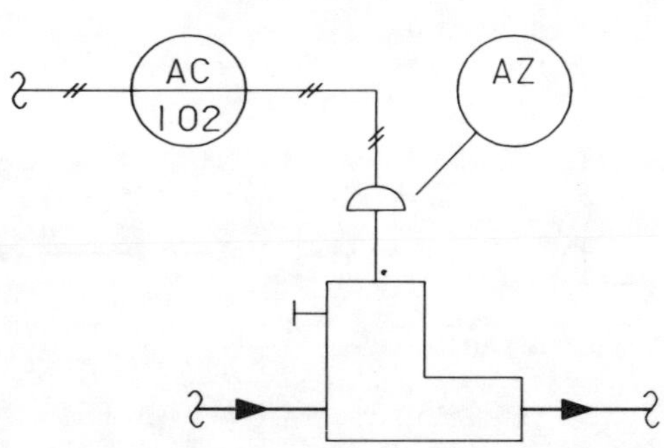

PROPORTIONING PUMP WITH AUTOMATIC AND MANUAL STROKE CONTROL</td>
<td>2
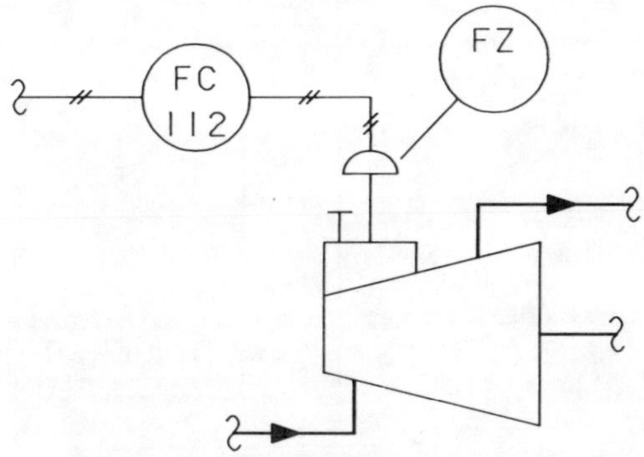

TURBINE WITH AUTOMATIC AND MANUAL SPEED CHANGER</td>
</tr>
</table>

6.10 EXAMPLES-MISCELLANEOUS COMBINATIONS

1

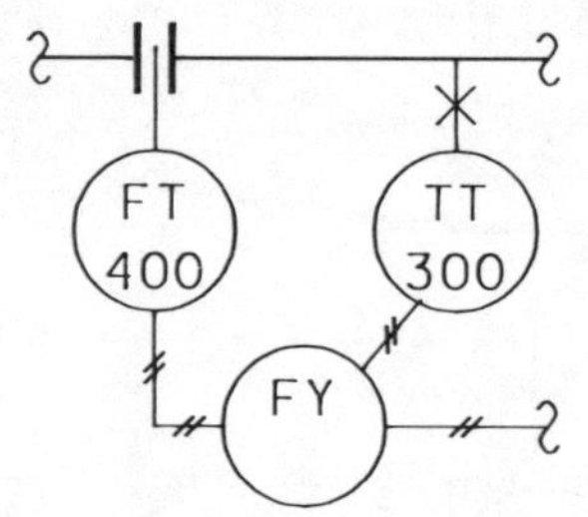

FLOW TRANSMITTER WITH EXTERNAL TEMPERATURE COMPENSATION

2

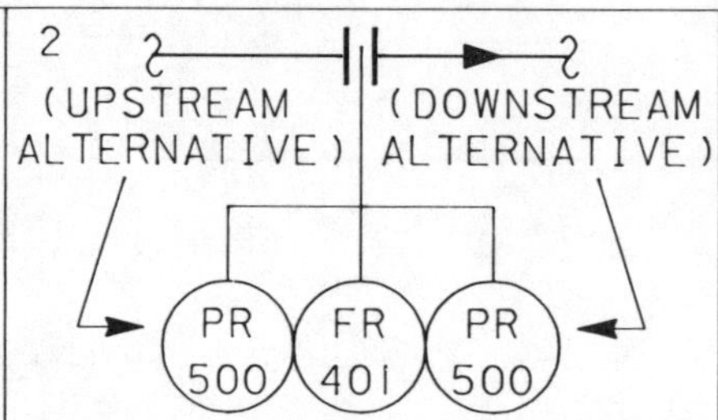

FLOW RECORDER WITH PRESSURE PEN CONNECTED TO TAP OF DIFFERENTIAL-PRESSURE-TYPE FLOW RECORDER

3

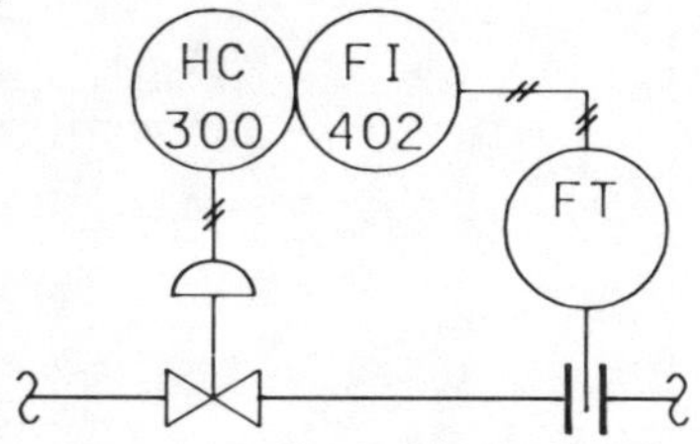

MANUAL LOADING STATION WITHOUT OUTPUT GAGE AND WITH FLOW RECEIVER INDICATOR

4

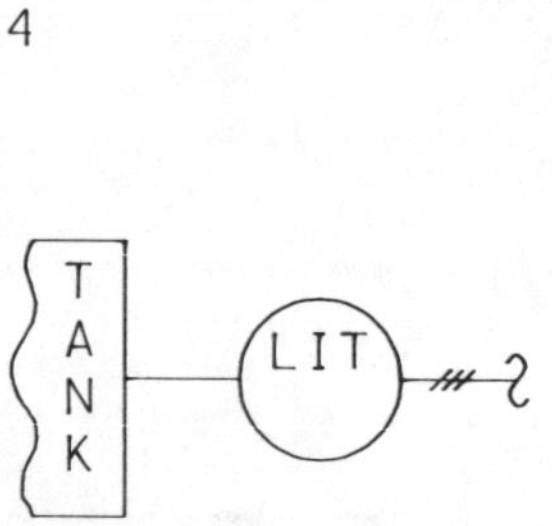

LEVEL TRANSMITTER WITH INTEGRAL INDICATION

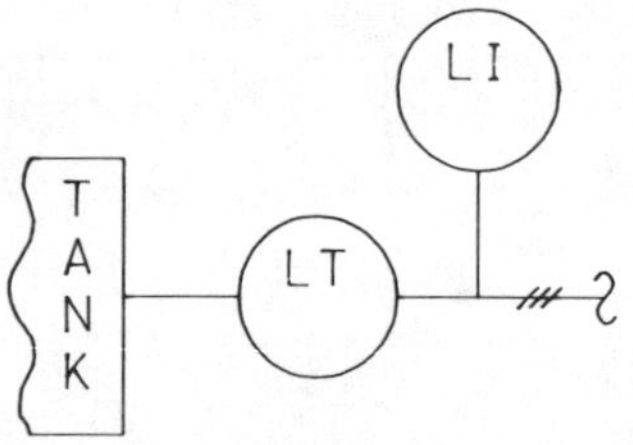

LEVEL TRANSMITTER WITH SEPARATE INDICATION

5

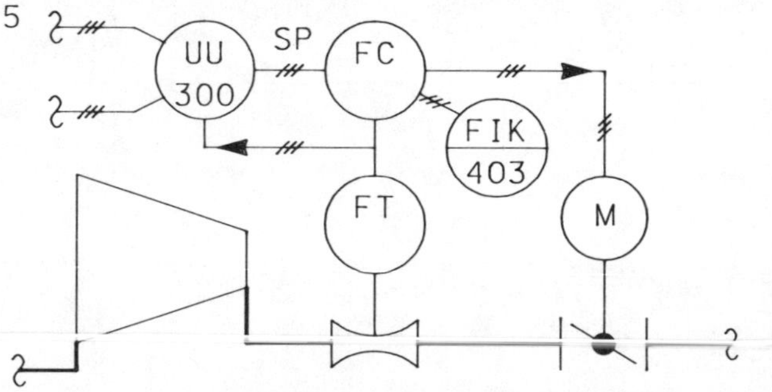

COMPLEX COMBUSTION CONTROL SYSTEM, UU-300, ADJUSTING SET-POINT OF FLOW CONTROL SYSTEM

6

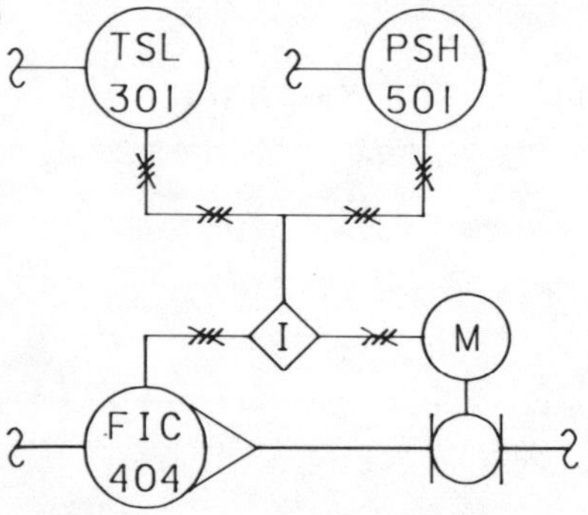

PREFERRED METHOD

TO BE USED IF INTERLOCK IS UNDEFINED OR COMPLEX. I-1 MAY BE SUBSTITUTED FOR I TO REFER TO INTERLOCK DETAIL NO. 1, TO BE DRAWN ELSEWHERE. THE OUTPUTS OF TSL-301 AND PSH-501 MAY ENTER THE INTERLOCK BOX SEPARATELY INSTEAD OF JOINTLY AS SHOWN

7

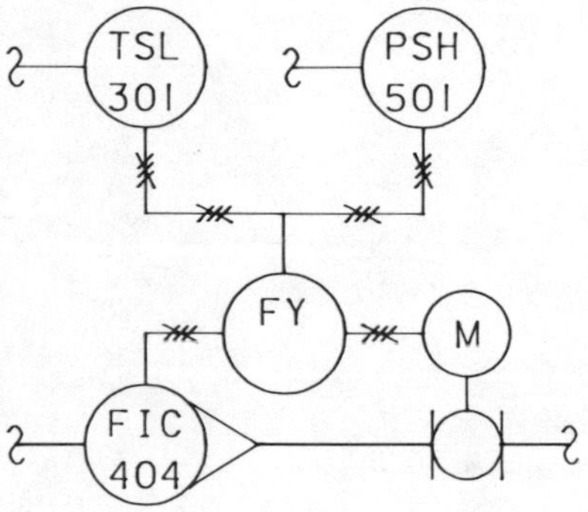

OPTIONAL METHOD

MAY BE USED IF RELAY EXISTS AND ONE DESIRES TO SHOW IT

FLOW CONTROL LOOP THAT MAY BE BLOCKED BY INTERLOCK ACTUATED BY EITHER LOW TEMPERATURE OR HIGH PRESSURE

6.10 EXAMPLES-MISCELLANEOUS COMBINATIONS (Contd.)

8

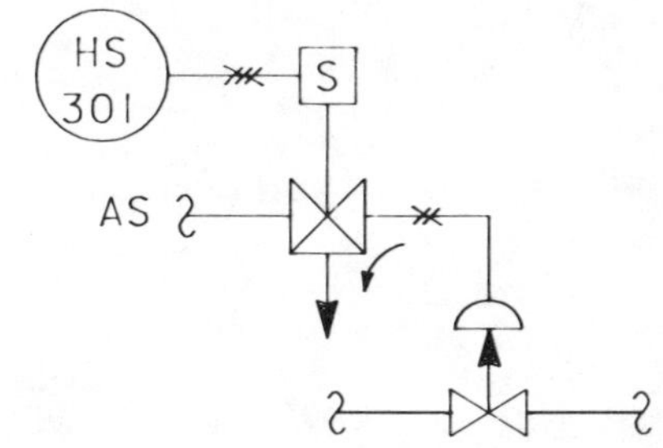

SOLENOID PILOT VALVE ACTUATED BY HAND SWITCH

9

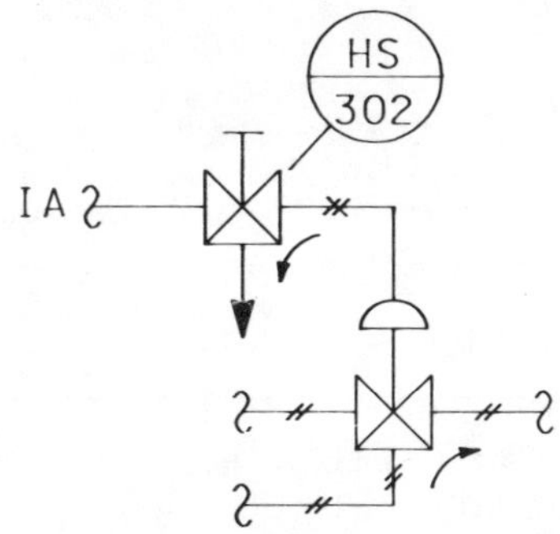

ALTERNATIVE 1

10

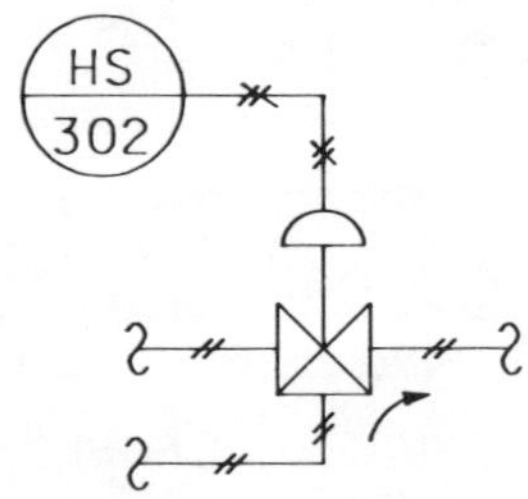

ALTERNATIVE 2

PNEUMATIC PILOT VALVE ACTUATED BY REMOTE HAND SWITCH

11

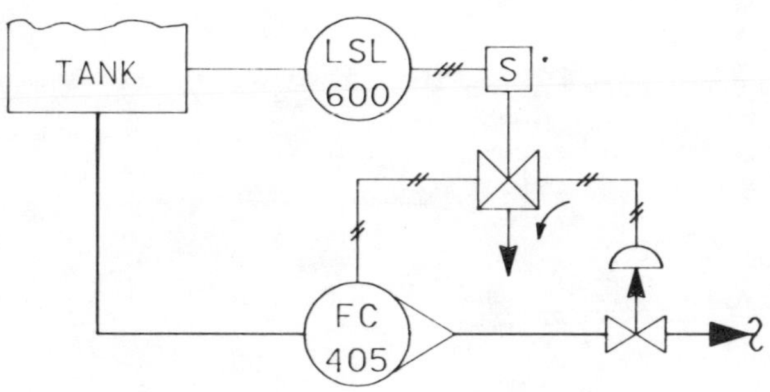

FLOW CONTROL LOOP WITH LOW-LEVEL INTERLOCK ACTUATED BY LEVEL SWITCH. THE SOLENOID MAY BE TAGGED FY-405

12

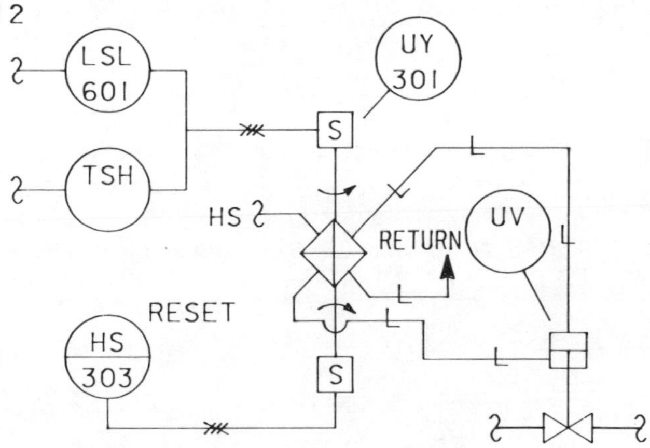

PISTON-ACTUATED ON-OFF VALVE TO CLOSE ON LOW LEVEL OR HIGH TEMPERATURE AND TO BE REOPENED BY PUSHBUTTON RESET OF DOUBLE-COIL SOLENOID VALVE

13

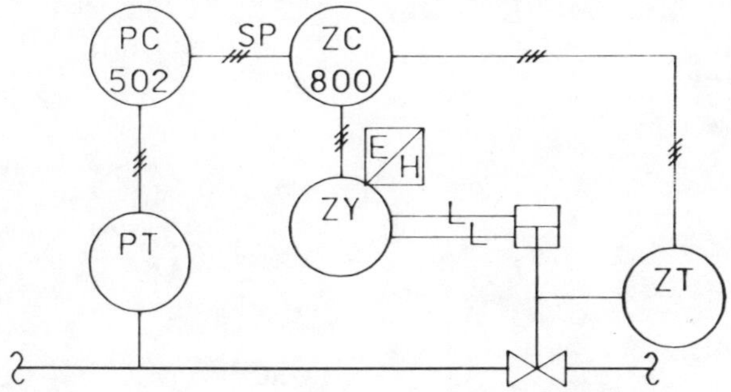

PRESSURE CONTROLLER ACTING THROUGH SERVO-VALVE ACTING AS VOLTAGE-TO-HYDRAULIC CONVERTER TO MODULATE DOUBLE-ACTING-CYLINDER-ACTUATED CONTROL VALVE HAVING POSITIONING SYSTEM

14

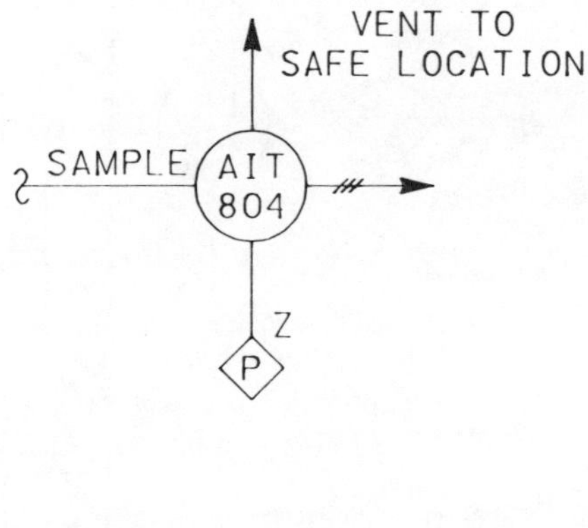

SAFETY PURGING (Z TYPE)

6.10 EXAMPLES-MISCELLANEOUS COMBINATIONS (Contd.)

15

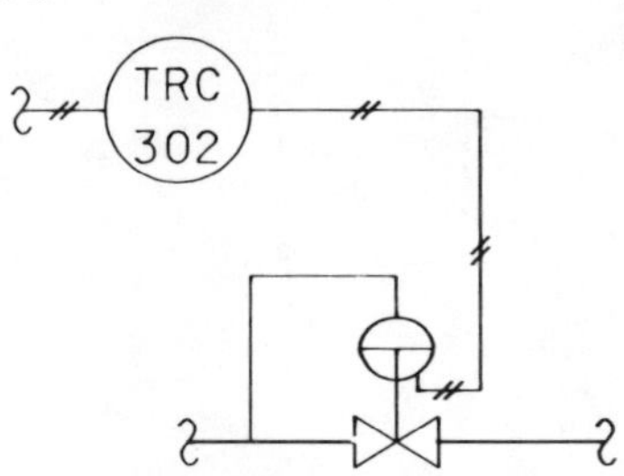

PRESSURE-BALANCED VALVE

16

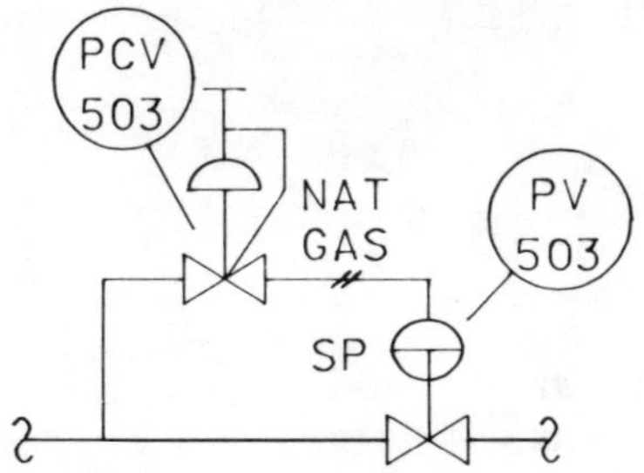

SELF-ACTUATED NATURAL GAS PRESSURE REGULATOR WITH REMOTE SET-POINT ADJUSTMENT

17

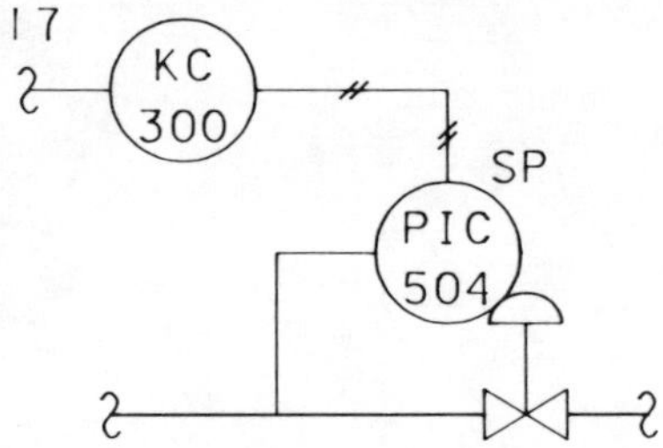

TIME-SCHEDULE CONTROLLER VARYING SET-POINT OF VALVE-MOUNTED PRESSURE CONTROLLER (TAG VALVE ASSEMBLY PIC/PV-504)

18

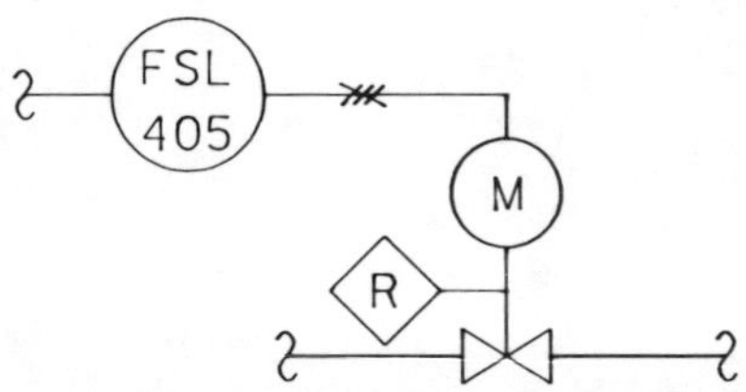

VALVE THAT UNLATCHES OR LATCHES WHEN TRIPPED AND REQUIRES HAND ACTUATION OF VALVE-MOUNTED RESET MECHANISM

19

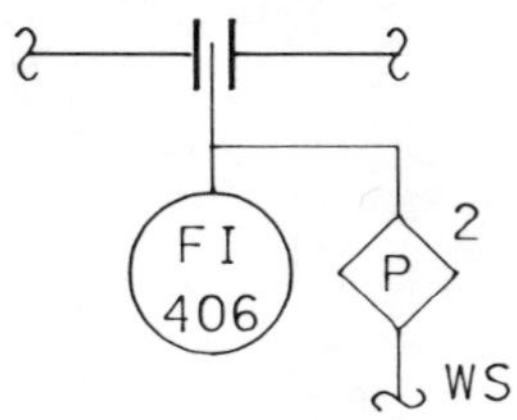

DIFFERENTIAL-PRESSURE TYPE FLOW INDICATOR WITH WATER PURGE* FOR BOTH LEADS (DENOTED BY NUMBER 2)

20

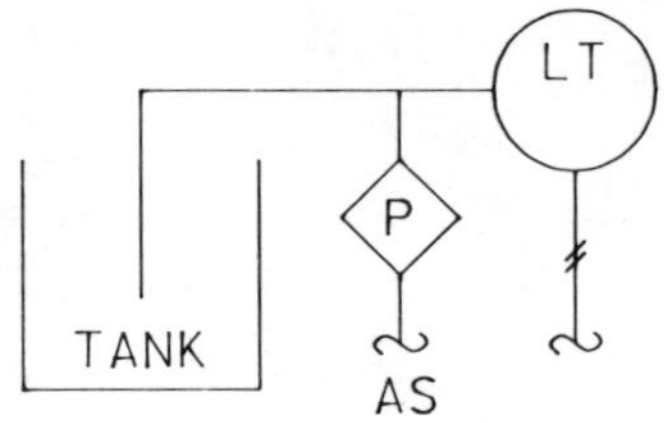

LEVEL TRANSMITTER FOR OPEN TANK WITH DIP TUBE AND AIR PURGE*

21

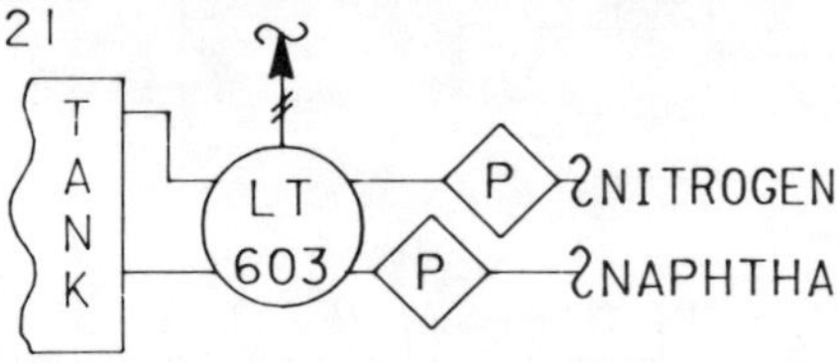

DIFFERENTIAL-PRESSURE TYPE LEVEL TRANSMITTER WITH GAS AND LIQUID PURGES*

22

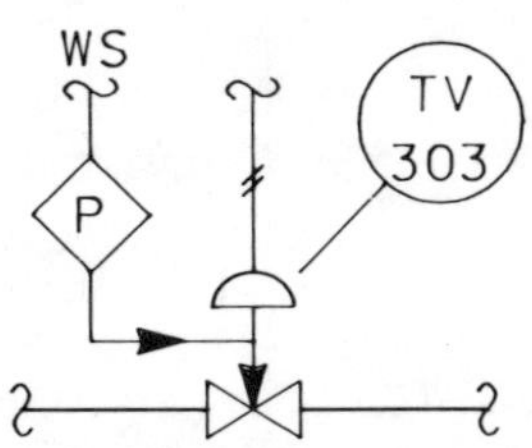

CONTROL VALVE WITH WATER-SEALED* STUFFING BOX. VALVE CLOSES ON AIR FAILURE

23

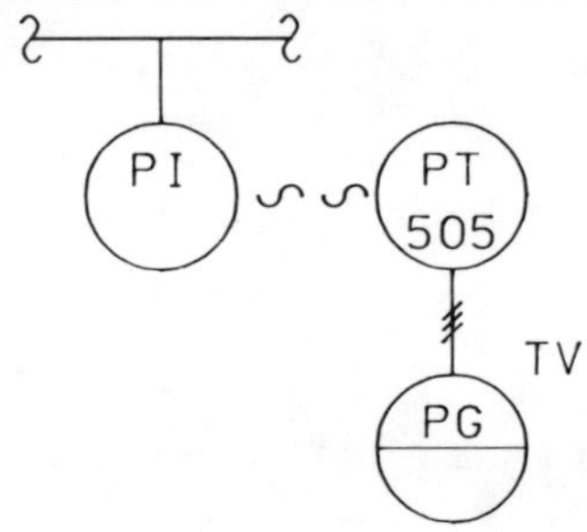

TELEVISION VIEWING OF PRESSURE GAGE

24

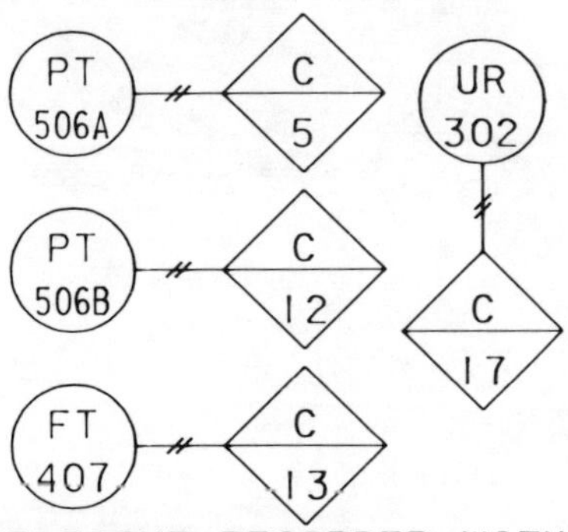

ONE-POINT RECORDER WITH PATCHBOARD INPUT CONNECTION

25

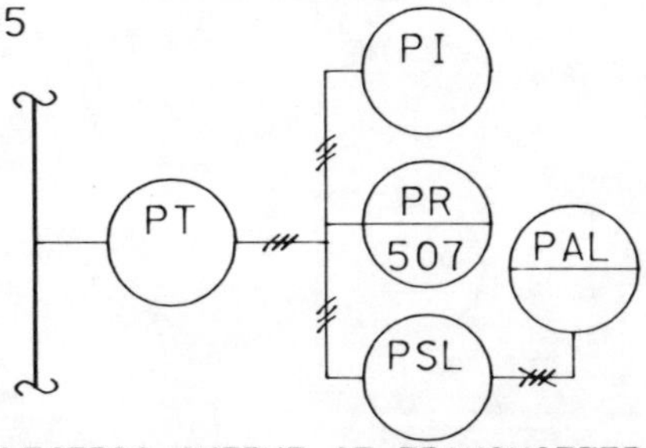

ELECTRIC OUTPUT OF TRANSMITTER CONNECTED TO MORE THAN ONE RECEIVER, REGARDLESS OF TYPE OF ELECTRIC SIGNAL

26

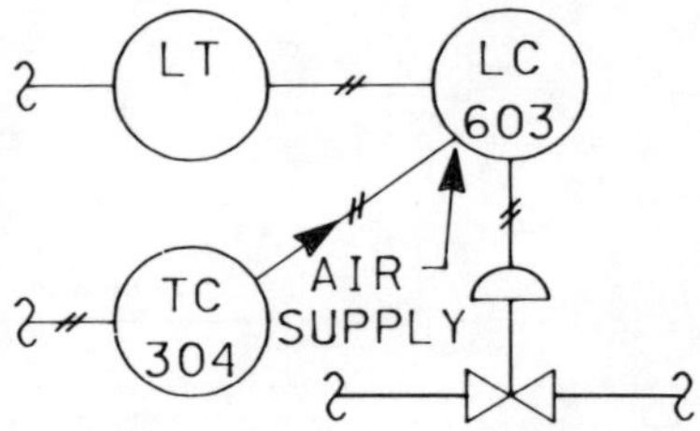

TEMPERATURE CONTROLLER OUTPUT ACTING AS AIR SUPPLY FOR LEVEL CONTROLLER

* The purge fluid supplies may use the same abbreviations as the instrument power supplies.

6.10 EXAMPLES-MISCELLANEOUS COMBINATIONS (Contd.)

27

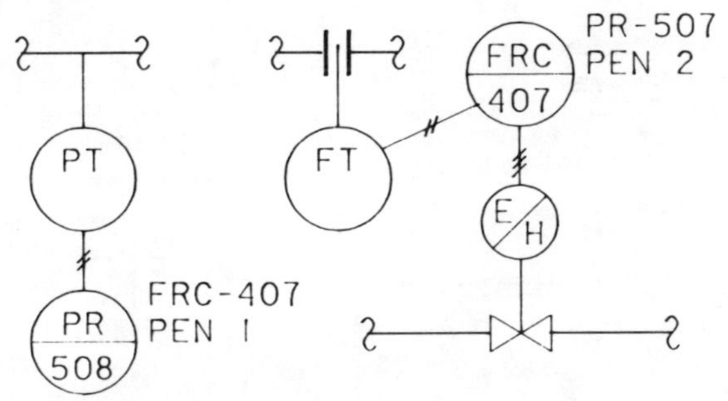

ALTERNATIVE 1

28

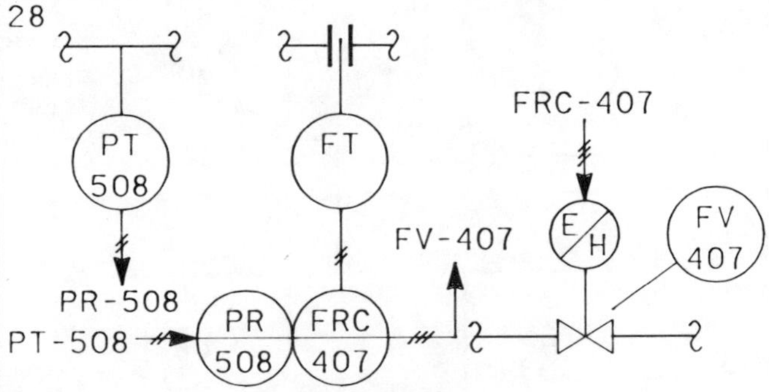

ALTERNATIVE 2

INTERRELATED INSTRUMENTS WHOSE SYMBOLS ARE SCATTERED ON DIAGRAM. (A MULTI-POINT INSTRUMENT, SUCH AS A DATA LOGGER, THAT IS DESIGNATED WITH POINT NUMBERS ON A DIAGRAM IS NOT EXPECTED TO HAVE THE SYMBOLS FOR THE VARIOUS POINTS TIED OR REFERENCED TOGETHER). PEN ASSIGNMENTS NEED NOT BE SHOWN ON A DIAGRAM IF IT IS THE USER'S PREFERENCE TO SHOW THEM IN AN INDEX

29

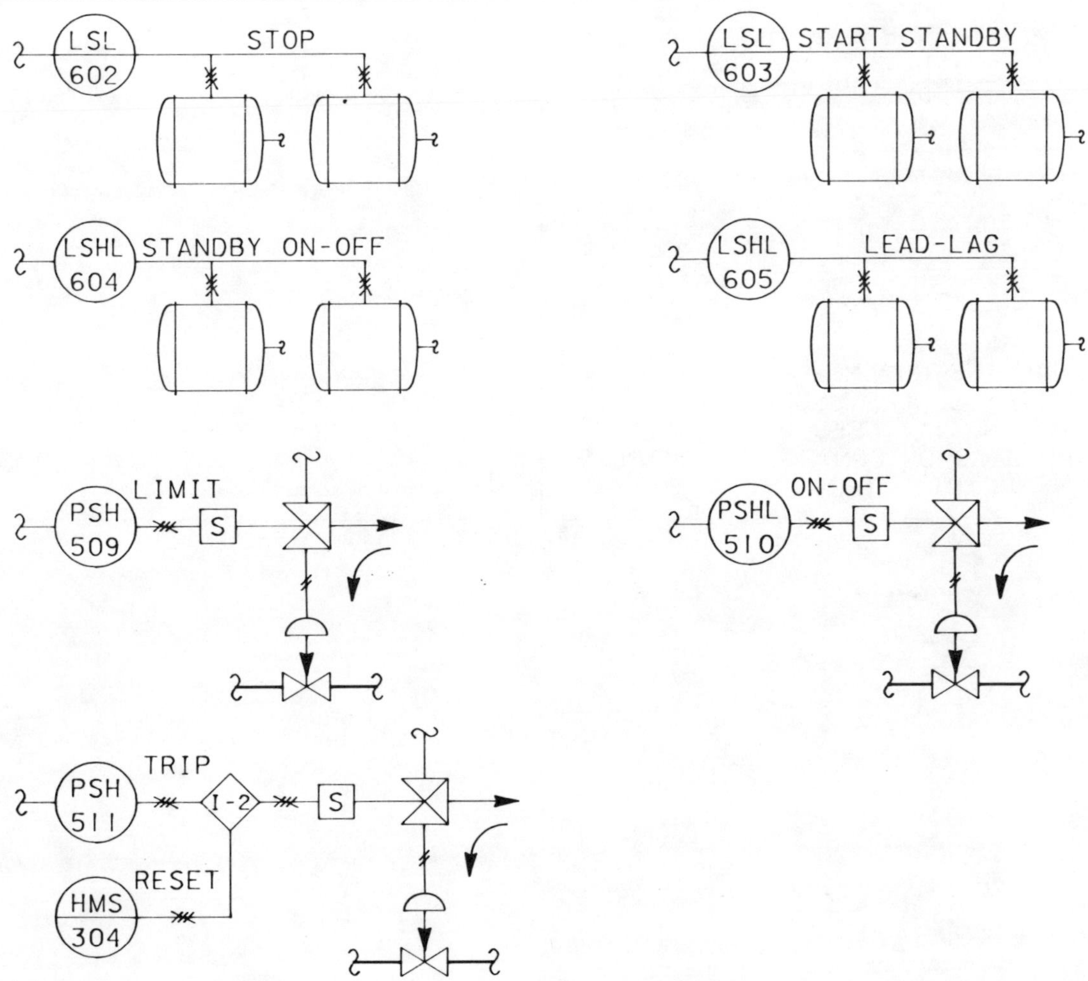

THE JUDICIOUS USE OF WORDS CAN CLARIFY DESIGN INTENT

6.10 EXAMPLES-MISCELLANEOUS COMBINATIONS (Contd.)

30

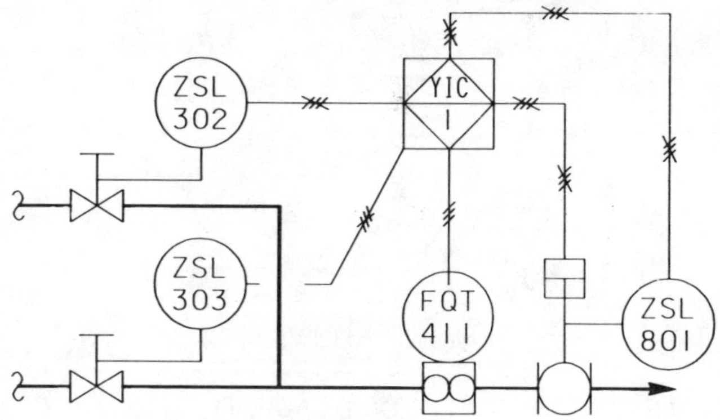

ALTERNATIVE 1

31

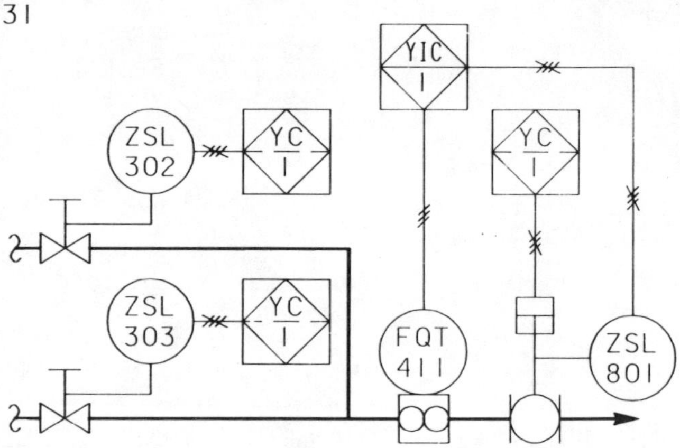

ALTERNATIVE 2

PROGRAMMABLE CONTROLLER SCHEME. ALTERNATIVE 2 ALLOWS MORE DEFINITION OF WHICH SIGNALS ARE ACCESSIBLE TO THE OPERATOR

32

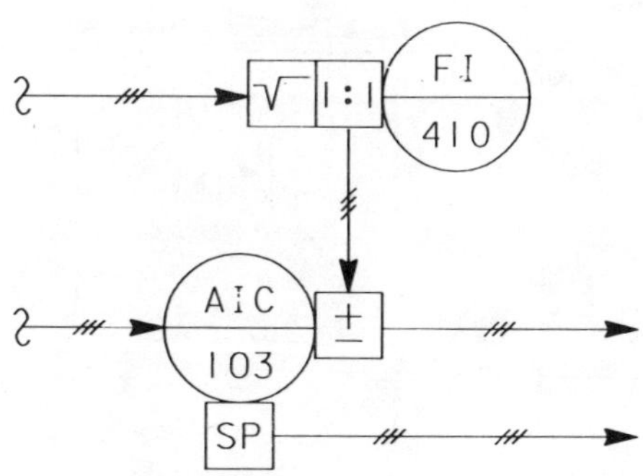

INDIVIDUAL INSTRUMENTS WITH INTEGRAL FUNCTION BLOCKS: SQUARE ROOT EXTRACTION, 1:1 FEED BOOST, FEED FORWARD BIAS ON CONTROLLER OUTPUT, SETPOINT OUTPUT SIGNAL

33

6.11 EXAMPLE - COMPLEX COMBINATIONS

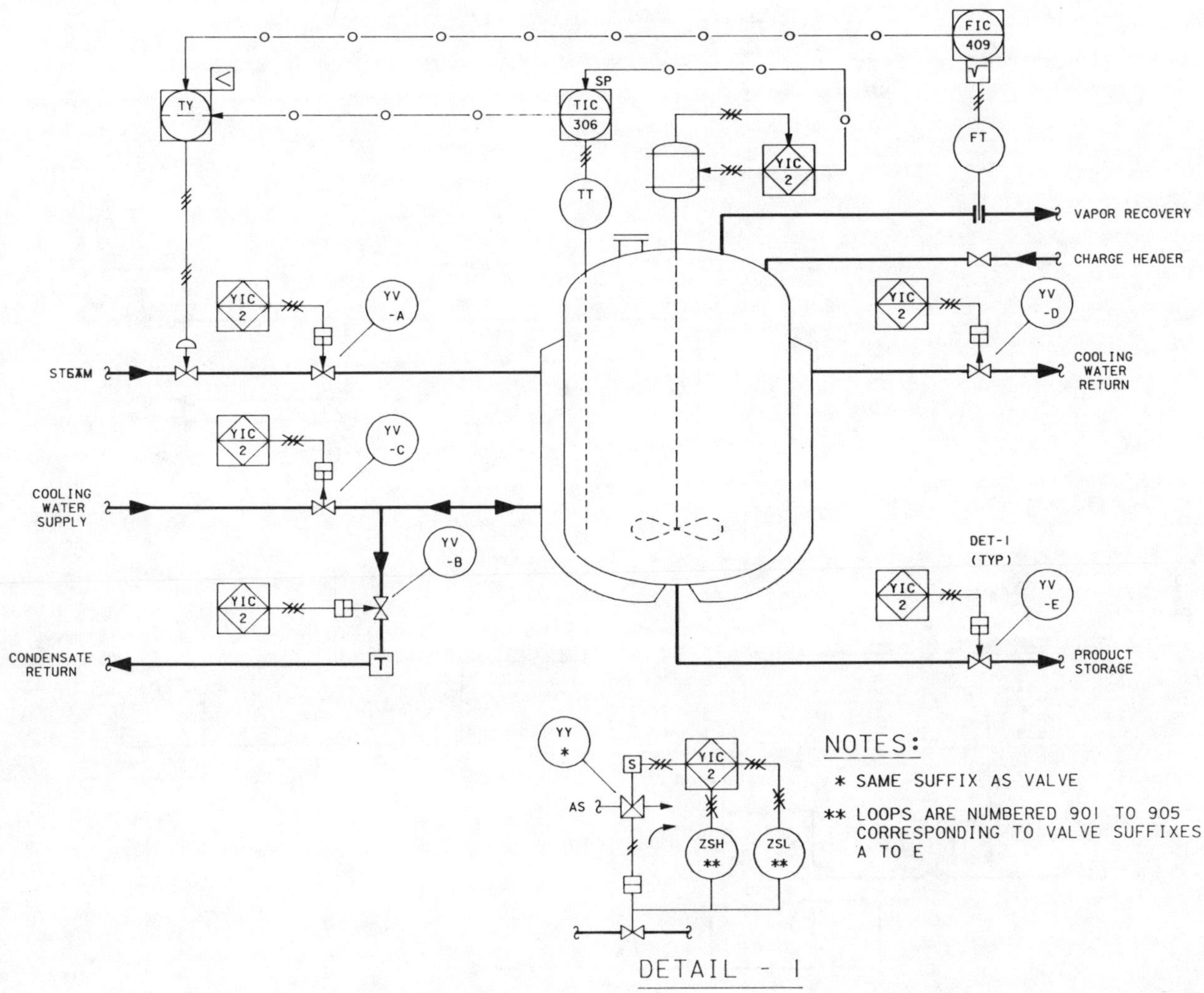

Example of complex batch kettle control involving both shared display/control and programmable logic control. The purpose of this drawing is to allow a general understanding of the control scheme and to define and identify the connected components. Detailed understanding would be obtained from the study of other documents.

6.12 EXAMPLE - DEGREE OF DETAIL *

I TYPICAL SYMBOLISM FOR SIMPLIFIED DIAGRAMS

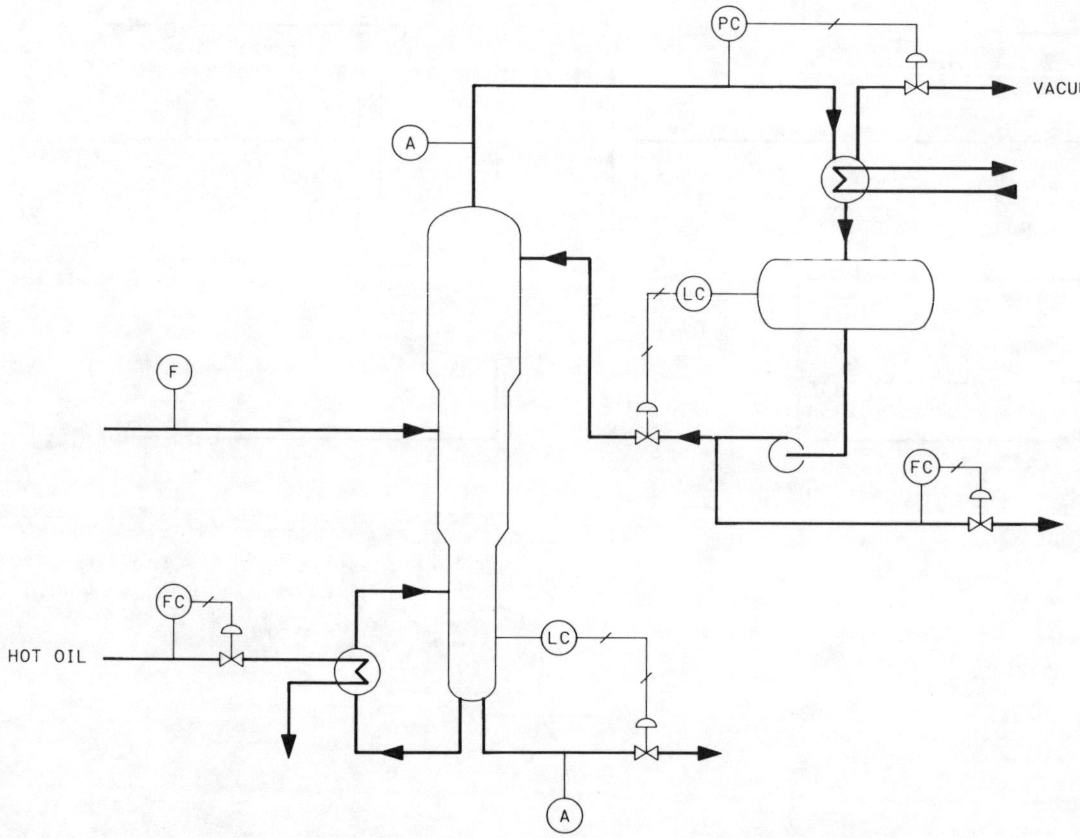

Simplified symbolism and abbreviated identification used to define the principal points of measurement and control interest.

* SEE SECTION 4.4 FOR DISCUSSION

6.12 EXAMPLE - DEGREE OF DETAIL *

2 TYPICAL SYMBOLISM FOR CONCEPTUAL DIAGRAMS

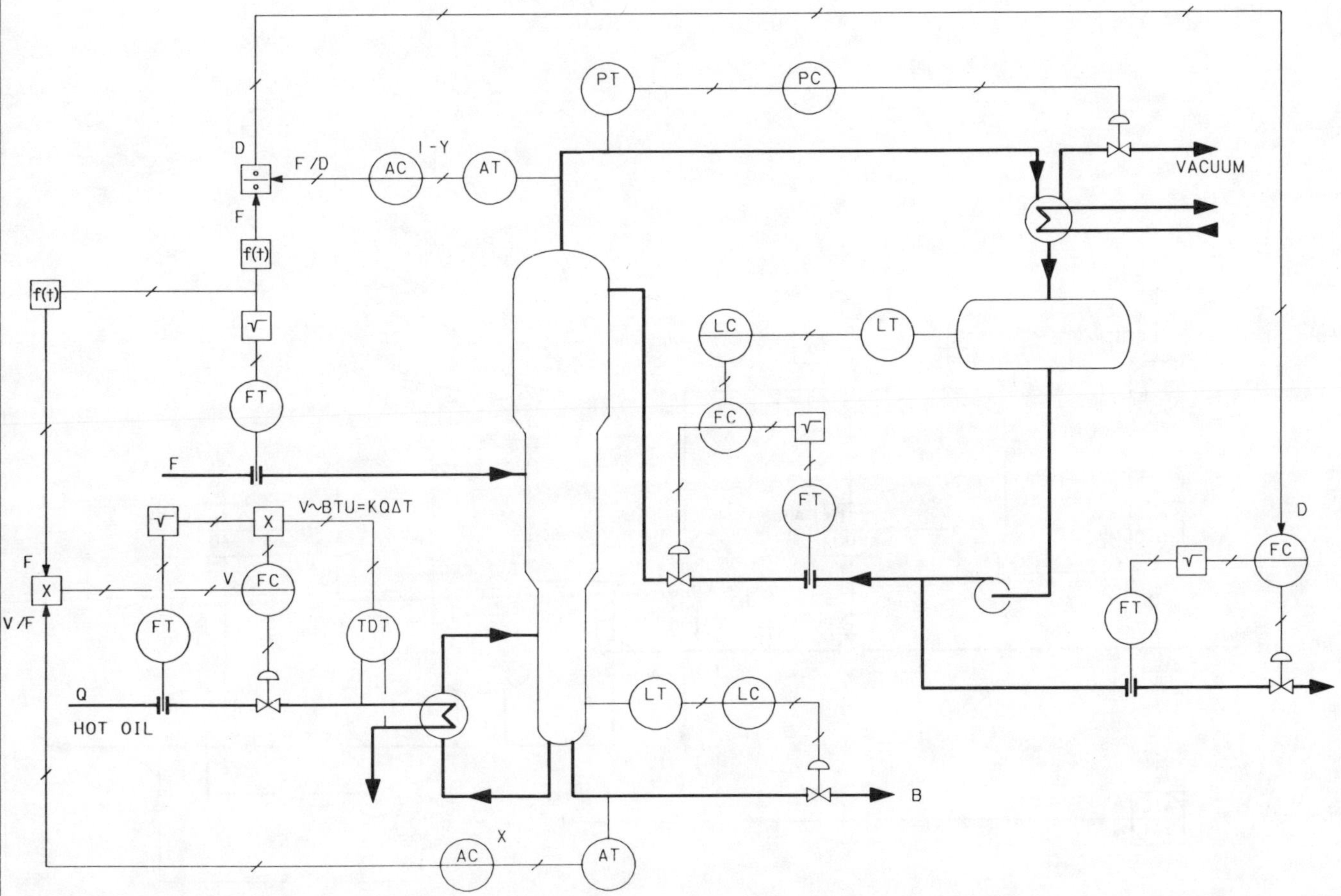

Functionally oriented symbolism and abbreviated identification used to develop control concepts without concern for specific hardware.

* SEE SECTION 4.4 FOR DISCUSSION

6.12 EXAMPLE - DEGREE OF DETAIL *

3 TYPICAL SYMBOLISM FOR DETAILED DIAGRAMS

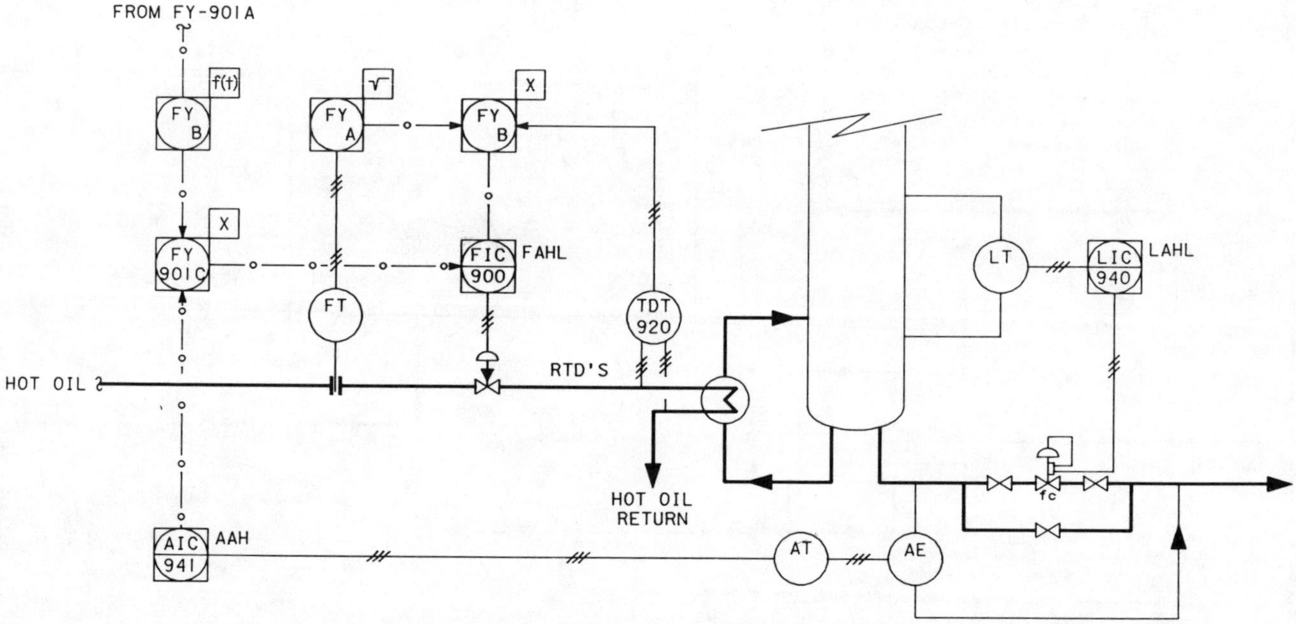

Detailed symbolism and more complete identification used to describe the control system when the type of hardware and kinds of signals have been chosen.

* SEE SECTION 4.4 FOR DISCUSSION

INDEX

Term	Para. No.	Page No.
-A-		
Accessible		
definition	3	9
graphic symbols	6.3	25
Actuator		
actuator action	6.7	33
graphic symbols	6.5	28
identification	5.1	15
Air Supply		
identification	6.2	24
Alarm		
definition	3	9
examples	6.9	44
identification	5.1	15
Analysis		
examples	6.8	34
identification	5.1	15,16
Assignable		
definition	3	10
Auto-Manual Station (Control Station)		
definition	3	10
identification	5.1	15
Auxiliary Location		
graphic symbol	6.3	25
Averaging		
graphic symbol	5.4	19
-B-		
Balloon, Bubble		
definition	3	10
graphic symbol	6.3	25
Behind-the-Panel		
definition	3	10
graphic symbol	6.3	25
Bias		
graphic symbol	5.4	21
Binary		
definition	3	10
graphic symbol	6.2	24
Board (Panel)		
definition	3	10
identification	6.3	25
Burner, Combustion		
example	6.8	34
identification	5.1	15
-C-		
Complex Combinations		
example	6.11	57
Computer Function		
graphic symbol	6.3	25
Computing Device		
definition	3	10
example	6.9	50
function	5.4	19
identification	5.1	15
Conceptual Diagrams		
example	6.12	59
Conductivity (see Analysis)		
example	6.8	34
Configurable		
definition	3	10
Controller		
definition	3	10
Control		
example	6.9	44
identification	5.1	15
Control Station (Auto-Manual Station)		
definition	3	10
example	6.9	46
identification	5.1	15
Control Valve		
definition	3	10
graphic symbols	6.4	27
identification	5.1	15
self-actuated	6.6	31
Converter		
definition	3	10
graphic symbols	5.4	21
Current (Electrical)		
graphic symbols	6.8	37
identification	5.1	15
-D-		
Damper		
graphic symbol	6.4	27
Data Link		
graphic symbol	6.2	24
Density		
example	6.8	34
Derivative		
graphic symbol	5.4	19
Detailed Diagrams		
example	6.12	60
Diaphragm Seal		
graphic symbol	6.3	26
Difference		
graphic symbol	5.4	19
Differential		
examples	6.6	31
	6.9	44
identification	5.1	15
Digital		
definition	3	10
graphic symbol	6.5	28
Dimension		
identification	5.1	15
Discrete Instruments		
graphic symbol	6.3	25
Distributed Control System		
definition	3	10
Dividing		
graphic symbol	5.4	20
Driver		
example	6.9	51
identification	5.1	15

Term	Para. No.	Page No.
-E-		
Electric Binary Signal		
graphic symbol	6.2	24
Electric Signal		
graphic symbols	6.2	24
Electric Supply		
identification	6.2	24
Electrohydraulic		
graphic symbol	6.5	29
Electromagnetic Signal		
graphic symbols	6.2	24
Event		
identification	5.1	15
Exponential		
graphic symbol	5.4	20
-F-		
Field Mounted (Local)		
graphic symbol	6.3	25
Final Control Element		
definition	3	10
identification	5.1	15
Flow Rate		
graphic symbols	6.8	35
identification	5.1	15
Force		
example	6.8	43
identification	5.1	15
Frequency		
identification	5.1	15
Function		
definition	3	10
examples	6.9	44-51
graphic symbols	6.3	25
identification	5.1	15
	5.4	19-22
Function Blocks		
identification	5.4	19-22
Fusible Plug		
example	6.6	32
-G-		
Gas Supply		
identification	6.2	24
Glass (Viewing Device)		
example	6.9	45
graphic symbols	6.8	38
identification	5.1	15
-H-		
Hand (Manually Actuated)		
example	6.9	45
graphic symbols	6.6	30
identification	5.1	15
Hand Wheel		
graphic symbol	6.5	29
High		
identification	5.1	15
High Limiting		
graphic symbol	5.4	21
High Selecting		
graphic symbol	5.4	20
Hydraulic Signal		
graphic symbol	6.2	24
Hydraulic Supply		
identification	6.2	24
-I-		
Identification		
combinations example	5.3	18
definition	3	10
letters	5.1	15
Indicate		
example	6.9	46
identification	5.1	15
Instrument		
definition	3	10
Instrument Air		
identification	6.2	24
Instrumentation		
definition	3	10
Instrument Line Symbols	6.2	24
Integral		
graphic symbol	5.4	19
Integrate		
example	6.9	47
identification	5.1	15
Interlock		
example	6.9	47,52,54
graphic symbol	6.3	26
-L-		
Letter Combinations	5.3	18
Level		
graphic symbol	6.8	38-39
	6.6	31
identification	5.1	15
Light		
example	6.9	47
graphic symbol	6.3	26
identification	5.1	15
Local (Field)		
definition	3	11
graphic symbol	6.3	25
Local Panel		
definition	3	11
identification	6.3	25
Loop		
definition	3	11
identification	4.3	13
Louver		
see Damper		
Low		
example	6.9	47
identification	5.1	15
Low Limiting		
graphic symbol	5.4	21
Low Selecting		
graphic symbol	5.4	21
-M-		
Manual Loading Station		
definition	3	11

Term	Para. No.	Page No.
Measurement		
definition	3	11
Mechanical Analysis		
identification	5.1	15
Mechanical Link		
graphic symbol	6.2	24
Middle (Intermediate)		
identification	5.1	15
Moisture		
see Analysis		
Momentary		
identification	5.1	15
Monitor		
definition	3	11
Multifunction		
example	6.9	49
identification	5.1	15
Multiplying		
graphic symbol	5.4	20
Multivariable		
graphic symbols	6.8	42
identification	5.1	15
-N-		
Nitrogen Supply		
identification	6.2	24
Non-Linear		
graphic symbol	5.4	20
-O-		
Orifice		
identification	5.1	15
Orifice Plate Taps		
graphic symbols	6.8	35
-P-		
Panel (Board) (Primary Location)		
definition	3	11
graphic symbols	6.3	25
Panel-Mounted (Board-Mounted)		
definition	3	11
graphic symbols	6.3	25
Pilot Light		
definition	3	11
graphic symbol	6.3	26
identification	5.1	16
Plant Air		
identification	6.2	24
Pneumatic Binary Signal		
graphic symbol	6.2	24
Pneumatic Signal		
graphic symbol	6.2	24
Point (Test)		
example	6.9	47
identification	5.1	15
Position		
graphic symbol	6.8	43
identification	5.1	15
Power		
graphic symbol	6.8	37
identification	5.1	15
Presence		
identification	5.1	15
Pressure Relief Valves		
see Safety Relief Valves		
Pressure (or Vacuum)		
graphic symbols	6.6	31-32
	6.8	39
identification	5.1	15
Print		
example	6.9	48
Primary Element (Sensor)		
definition	see Sensor	
graphic symbols	6.8	34-43
Process		
definition	3	11
Process Variable		
definition	3	11
Program		
definition	3	11
Programmable Logic Controller		
definition	3	11
example	6.10	56
	6.11	57
graphic symbol	6.3	25
Proportional		
graphic symbol	5.4	19
Purge		
example	6.10	54
graphic symbol	6.3	26
-Q-		
Quantity		
examples	6.9	47
graphic symbols	6.8	40
identification	5.1	15
-R-		
Radiation		
graphic symbol	6.8	40
Ratio		
example	6.9	45
identification	5.1	15
Record		
example	6.9	48
identification	5.1	15
Relay		
definition	3	11
examples	6.9	50-51
graphic symbols	5.4	21
Relief Valve		
see Safety Relief Valve		
Reset		
graphic symbol	6.3	26
Restriction Orifice		
graphic symbol	6.6	30
Reverse Proportional		
graphic symbol	5.4	21
Root Extraction		
graphic symbol	5.4	20
Rupture Disc		
graphic symbols	6.6	32

Term	Para. No.	Page No.
-S-		
Safety		
identification	5.1	15,16
Safety Purging		
example	6.10	53
Safety Relief Valves		
graphic symbols	6.6	31,32
Scan		
definition	3	11
example	6.9	46
identification	5.1	15
Self-Actuated Devices		
graphic symbols	6.6	30-32
identification	4.2.5	12
Sensor (Primary Element)		
definition	3	11
graphic symbols	6.8	34-42
Set Point		
definition	3	11
Shared Controller		
definition	3	11
graphic symbols	6.3	25
Shared Display		
definition	3	11
graphic symbols	6.3	25
Sight Glass		
graphic symbol	6.6	30
Signal Monitor		
graphic symbol	5.4	22
Simplified Diagram		
example	6.12	58
Solenoid		
graphic symbol	6.5	28,29
Specific Gravity		
example	6.8	34
Speed		
identification	5.1	15
Steam Supply		
identification	6.2	24
Straightening Vane		
graphic symbol	6.6	30
Summing		
graphic symbol	5.4	19
Switch		
definition	3	11
example	6.9	48
identification	5.1	15
-T-		
Temperature		
graphic symbols	6.8	40-41
identification	5.1	15
Test Point		
definition	3	11
example	6.9	47
identification	5.1	15
Time Function		
graphic symbol	5.4	20
Time Rate of Change		
identification	5.1	15
Time, Time Schedule		
graphic symbols	6.8	37
identification	5.1	15
Totalize		
see Integrate		
Transducer		
definition	3	12
Transmitter		
definition	3	12
example	6.9	49
identification	5.1	15
Traps		
example	6.6	32
Typical Letter Combinations	5.3	18
-U-		
Unclassified		
identification	5.1	15
Undefined Signal		
graphic symbols	6.2	24
User's Choice		
examples	6.8	34,49
identification	5.1	15
-V-		
Vacuum		
see Pressure		
Valve		
see Control Valve		
Velocity Limiter		
graphic symbol	5.4	21
Vibration		
identification	5.1	15
Voltage		
graphic symbols	6.8	35
identification	5.1	15
-W-		
Water Supply		
identification	6.2	24
Weight		
identification	5.1	15
Well		
graphic symbols	6.8	40,41
identification	5.1	15
-X-		
X-Axis		
identification	5.1	15
-Y-		
Y-Axis		
identification	5.1	15
-Z-		
Z-Axis		
identification	5.1	15